T0387743

Geographies of Transport and Ageing

Angela Curl · Charles Musselwhite
Editors

Geographies of Transport and Ageing

palgrave
macmillan

Editors
Angela Curl
Department of Geography
University of Canterbury
Christchurch, New Zealand

Charles Musselwhite
Centre for Innovative Ageing
Swansea University
Swansea, UK

ISBN 978-3-319-76359-0 ISBN 978-3-319-76360-6 (eBook)
https://doi.org/10.1007/978-3-319-76360-6

Library of Congress Control Number: 2018934656

© The Editor(s) (if applicable) and The Author(s) 2018
This work is subject to copyright. All rights are solely and exclusively licensed by the Publisher, whether the whole or part of the material is concerned, specifically the rights of translation, reprinting, reuse of illustrations, recitation, broadcasting, reproduction on microfilms or in any other physical way, and transmission or information storage and retrieval, electronic adaptation, computer software, or by similar or dissimilar methodology now known or hereafter developed.
The use of general descriptive names, registered names, trademarks, service marks, etc. in this publication does not imply, even in the absence of a specific statement, that such names are exempt from the relevant protective laws and regulations and therefore free for general use.
The publisher, the authors and the editors are safe to assume that the advice and information in this book are believed to be true and accurate at the date of publication. Neither the publisher nor the authors or the editors give a warranty, express or implied, with respect to the material contained herein or for any errors or omissions that may have been made. The publisher remains neutral with regard to jurisdictional claims in published maps and institutional affiliations.

Cover credit: Hero Images/Getty

Printed on acid-free paper

This Palgrave Macmillan imprint is published by the registered company Springer International Publishing AG part of Springer Nature
The registered company address is: Gewerbestrasse 11, 6330 Cham, Switzerland

Angela Curl would like to thank the supportive people she has worked with, across different fields of study with a common focus on people and transport, for the opportunities they have given her to be involved in wide-ranging and interesting research projects. She would like to thank her family and friends for ongoing support and to nan: for giving me my first insight into what the implications of the built environment can be for mobility, health and well-being.

Charles Musselwhite would like to thank all the inspirational people he has worked with over his career that have given him space to develop his thinking and research. He is very thankful to his family, Claire, William and Art for giving him time and space for developing his academic career and to his mum and dad and brother James for always inspiring and supporting him.

Preface

This book presents a collection of geographical research on the intersection of ageing, mobility and transport. It is not really possible to think about many of the challenges facing ageing societies worldwide without thinking about interactions between people and place and the mobility and transport which enable this interaction. Despite this, policy often maintains extending independence and ageing in place as vital requirements to a dignified healthy later life without really addressing issues of mobility.

When thinking about transport, travel and mobilities of an ageing population, it is impossible to do so without taking a multi-disciplinary approach. Naturally mobility involves geography, the movement of people over space and time, and involves cultural, social and psychological elements. Transport is essentially a means to overcome the geography; the distance between people and place. Yet mobility is not just literal but also virtual, social and cultural. We address this for an ageing population, bringing a critical geographical and gerontological approach. In this edition, we emphasise the importance of different modalities, of different places, spaces and cultures, of technologies and cast an eye to the future. We have separated the book into sections based on the focus of the chapters contained within them, but studies of geography, ageing and transport needs to bring them together to make a difference to policy and practice.

The origins of this book lie in a session at the 2015 Royal Geographical Society with the Institute of British Geographers Annual International Conference, organised by the editors and sponsored by the *Transport Geography Research Group*. The session was called *Maintaining Mobility: Geographies of Transport and Ageing*. We are grateful to Palgrave Macmillan for showing such interest in the topic and encouragement and support for pulling together this edited collection following the conference. The idea for the session came from the same conference the previous year, and a series of sessions held on 'Urban and Suburban Geographies of Ageing' which were sponsored by the *Urban Geography Research Group and the Geography of Health Research Group (now the Geographies of Health and Wellbeing Research Group)*. Mobility and transport were key themes throughout these sessions, despite transport not being an explicit focus for the session, supporting Shaw and Sidaway's (2011) assertion that transport is core to much of human geography without being identified as such. Indeed when thinking about ageing in urban, suburban and rural places, issues of transport and mobility cannot be ignored. The discovery that there had never been a session focussed explicitly on transport and ageing led to the organisation of our session the following year and subsequently this edited collected to draw together geographical research on ageing, transport and mobilities. We would like to thank the *Transport Geography Research Group* and the Royal Geographical Society for their support in organising this session and to the presenters who contributed to the debate. Most of all we thank all the contributors for their contributions to this book.

Christchurch, New Zealand — Angela Curl
Swansea, UK — Charles Musselwhite

Contents

Editors and Contributors

About the Editors

Dr. Angela Curl is a Lecturer in health geography at the University of Canterbury. Angela's research interests are focused around understanding people's perceptions and experiences of transport and mobility and how these intersect with the built environment, with a particular focus on older people's mobilities as a key interface between research in transport and health. She has previously worked in urban studies at the University of Glasgow, the OPENspace research centre at the University of Edinburgh and as a transport planning practitioner.

Dr. Charles Musselwhite is Associate Professor in gerontology, Centre for Innovative Ageing (CIA) at Swansea University. Charles heads-up the group's environments and ageing research strand. His research examines mobility and ageing, specifically studying travel behaviour, road user safety and relationships with health and well-being. Prior to joining Swansea, he worked as a Senior Lecturer at the Centre for Transport Studies, University of the West of England.

Contributors

Dr. Julie Clark lectures in sociology and social policy at the University of the West of Scotland and is an Associate Director of the ESRC Doctoral Training Partnership at the Scottish Graduate School for Social Science. Her field of interest is the relationship between policy, health and well-being, with particular reference to the role of transport and social inclusion. Julie's applied, policy-relevant research has been recognised with a Research Council UK Award for Research Impact and a Sir Peter Hall Award for Wider Engagement.

Dr. Margaret Currie is a human geographer, interested in how different types of spaces affect health and well-being of people, more specifically how being in a space can affect people—both positively and negatively—and the ways in which interventions can impact on this. Rural resilience and innovative service provision have been a focus of her research, particularly transport, health services and the use of ICTs in enabling new forms of accessibility in rural areas, also the ways in which rural people and communities can adapt or become resilient to the emergency and everyday changes within communities.

Prof. Benedicte Deforche is Full Professor at the Department of Public Health (Faculty of Medicine and Health Sciences, Ghent University). She is head of the Research Group Health Promotion and of the master programme in Health Education and Promotion. She is president of the work group "nutrition and physical activity" within the Flemish Health Prevention Policy Agency. She is also member of the General Assembly and member of the Board of Directors of the Flemish Institute for Healthy Living. She is Fellow of the International Society for Behavioral Nutrition and Physical Activity and member of the editorial board of the International Journal of Behavioral Nutrition and Physical Activity.

Prof. Bas de Geus is Assistant Professor at the Human Physiology Research Group (MFYS) at the Vrije Universiteit Brussel. He studies the health impact of active mobility and in particular cycling for transportation. His studies cover a wide variety of research topic as: impact of cycling for transportation, as a form of physical activity on different health parameters, effect of air pollution on health and characteristics of bicycle crashes in different age groups. Bas is a board member of the Scientists for Cycling network of the European Cyclists' Federation.

Dr. Gillian Dowds is a Research Fellow in the Department of Geography and Environment at the University of Aberdeen. She is currently examining the long-term impacts of recent flooding in North-East Scotland. Her main research interests include ageing in rural areas and well-being. Her Ph.D. adopted interdisciplinary methods, in the investigation of informing, designing then evaluating appropriate technology for largely housebound older adults living in rural areas, in order to enhance opportunities for engaging with their community.

Dr. Helen Fitt is a Postdoctoral Fellow at the University of Canterbury in Christchurch, New Zealand. She has long-standing interests in innovative uses of research methods and in social and cultural aspects of transport. She is currently working on a New Zealand National Science Challenge-funded project exploring the implications for older people and ageing societies of a transition to autonomous vehicles. Her aim is to try to encourage positive outcomes from such a transition, especially in terms of social equity.

Mark Gorman is Senior Policy Advisor at HelpAge International. He joined HelpAge in 1988 and was Deputy Chief Executive between 1991 and 2007, after which he continued to head the policy and advocacy team. Most recently, his work has focussed on the development of HelpAge's organisational strategy and on issues of ageing and health. He is also an Institutional Fellow at the Oxford Institute of Population Ageing. He has a degree in history from the University of Cambridge, and master's degrees from Cambridge and Bristol Universities. He has published a number of articles on aspects of ageing in low- and middle-income countries. He was awarded an MBE in 2008.

Dhruvi Kothari is an aspiring sustainable infrastructure and transportation planner at University of Pennsylvania, USA. She did her undergraduate degree in urban planning from CEPT University, India, and since has been actively participating in mobility design and planning field.

Prof. Judith Masthoff is a Chair in computing science at the University of Aberdeen. Her research interests cover areas including intelligent user interfaces, personalisation, digital behaviour interventions, recommender systems, evaluation of adaptive systems, e-health, sustainable transport and intelligent tutoring systems. She is the Editor in Chief of the User Modeling and User-Adapted Interaction journal and a Director of User Modeling Inc., the professional association of user modelling researchers. She is also Dean of the University Postgraduate Research School.

Shauna McGee is a doctoral student in clinical psychology at the University of Zürich, Switzerland. Her research interests focus on healthy ageing and the psychological, biological, social and environmental aspects of health and ageing processes. With a background in health and clinical psychology, she has worked in healthy ageing research in the UK and internationally. Her previous research includes environmental and design factors affecting health and well-being and the consideration of ageing in policy and governance models. She also contributed to the *Future of Ageing* Foresight report (2015) on the environment and mobility needs of an ageing population.

Dr. Jon Minton is AQMeN Research Fellow in the College of Social Sciences, University of Glasgow. Jon's main interests are in population health and the determinants of health inequalities between social groups, with a particular focus on the mortality effects of socioeconomic deprivation in Scotland. He has pioneered the use of Lexis surfaces—data maps of social and health outcomes that vary with both age and time—beyond their original applications in demography, to addressing and exploring and addressing a broader range of epidemiological and sociological challenges.

Dr. Talat Munshi is working as PostDoc UNEP-DTU Partnership, Denmark Technical University, and is also an Associate Professor at the Faculty of Planning, CEPT University, India. Prior to his current employment, he has worked as Lecturer in Transport at Faculty of ITC (University of Twente), at TERI University in Delhi, India, and as Lecturer at Faculty of Planning, CEPT University. His Ph.D. is from University of Twente and has two master's degrees: in planning, from CEPT University, and in urban infrastructure management from Faculty of ITC, University of Twente, the Netherlands.

Dr. Lorna Philip is Senior Lecturer in human geography and Deputy Head of the School of Geosciences at the University of Aberdeen. Her research focuses on social inclusion and demographic ageing in rural areas. Recent research projects, funded by UK Research Councils, have focused on retirement transition migration into rural areas, the potential of new digital technologies to promote social interaction among older rural adults, patterns of rural digital exclusion and explorations of rural digital divides.

Dr. Kate Pangbourne holds an EPSRC/LWEC personal fellowship award at the Institute for Transport Studies, University of Leeds, UK. She has an MA (Hons) in philosophy with English literature, an MSc in sustainable rural development and a Ph.D. in geography (environmental) focused on Scottish transport governance and spent 6 years as a postdoctoral researcher at the University of Aberdeen (UK). She has a highly interdisciplinary background, encompassing environmental sustainability, transport geography, rural development, technology and human–computer interaction, political/social science and philosophy, generally addressing complex problems through changing individual attitudes, behaviours and practices or through governance design.

Prof. Judith Phillips is Deputy Principal (Research) at the University of Stirling and Professor of Gerontology. Her research interests are in the social, behavioural and environmental aspects of ageing. A geographer by background she has a long-term interest in aspects of transport and ageing, space and place and design of urban environments for older people. She has published widely in social policy, social work, environmental gerontology and social care and has contributed with colleagues to the Foresight Report 2015 on aspects of mobility and environment. Her publications include *Planning and design of ageing communities* in Skinner, M. et al. *Geographical Gerontology* (Routledge, 2017).

Prof. Gina Porter has led a series of mobility-related research projects in sub-Saharan Africa over the last 30+ years and published widely in this field. Her latest book, *Young People's Daily Mobilities in Sub-Saharan Africa*, was published by Palgrave in 2017. Uneven power relationships and associated issues of exclusion are linking themes through her work, much of which has a strong gender component. Associated with this is a focus on developing innovative methodologies for effective field research.

Midhun Sankar is an urban transport planner working at Urban Mass Transit Company, India. He has a degree in architecture from School of Planning and Architecture, Bhopal, and postgraduate degree in urban planning with specialisation in Transport from CEPT University, Ahmedabad. He is keen on studying impacts of the urban transport on society especially disadvantaged groups.

Amleset Tewodros is a development and humanitarian worker with over two decades of work experience in addressing the needs of older people in low-income settings mainly working in East Africa. Amleset has been part of research programmes funded by the AFCAP and has co-authored three successive research projects on mobility and transport needs of older people and on gender mainstreaming in the transport policies of Tanzania and how these benefit women experiencing intersectional vulnerability due to age, disability and gender. Amleset is currently the Head of Programmes for the Africa Region of HelpAge International and oversees operations in seven Eastern and Southern African countries.

Dr. Sara Tilley is a postdoctoral researcher at the OPENspace research centre at the University of Edinburgh. She is a transport geographer, interested in exploring the links between mobility in the urban environment and health and well-being, especially in younger and older age. Her Ph.D. in ageing and mobility at the University of St Andrews was funded by a CASE Studentship awarded by the ESRC in partnership with the MMM Group (formerly MRC McLean Hazel), a leading transport consultancy. She has previously worked in the public and private sectors.

Dr. Jelle Van Cauwenberg is a postdoctoral research fellow at the Department of Public Health of Ghent University. In 2015, he defended his Ph.D. focusing on the environmental determinants of physical activity among older adults. With his postdoctoral research project, he aims to examine the contribution of e-bikes to older adults' physical activity, mobility and social participation. In addition, he is engaged in several research projects examining the environmental determinants of physical activity and sedentary behaviour in different age groups.

List of Figures

Part I

Context

1

Geographical Perspectives on Transport and Ageing

Charles Musselwhite and Angela Curl

Introduction

In terms of ageing, we are living in unprecedented times. People across the globe are living longer than ever before, and societies are ageing at increasing rates. In low- to middle-income countries, reductions in mortality at young ages have fuelled this growth. A person born today in Brazil, for example, can expect to live 20 years longer than someone born 50 years ago (WHO 2015). For the first time, life expectancy across the globe is over 60 years of age. In high-income countries, someone born now can expect to live up to around 80 years of age on

C. Musselwhite (✉)
Centre for Innovative Ageing, College of Human and Health Sciences, Swansea University, Swansea, UK
e-mail: c.b.a.musselwhite@swansea.ac.uk

A. Curl
Department of Geography, University of Canterbury—Te Whare Wānanga O Waitaha, Christchurch, New Zealand
e-mail: angela.curl@canterbury.ac.nz

© The Author(s) 2018
A. Curl and C. Musselwhite (eds.), *Geographies of Transport and Ageing*,
https://doi.org/10.1007/978-3-319-76360-6_1

average (ONS 2015). There are not simply a growing number of older people, but also a growing proportion of older people due to people living longer and declining birth rates in many countries. Across Europe, for example, people aged over 65 years will account for 29.5% of the population in 2060 compared to around 19% now (EUROSTAT 2017). The share of those aged 80 years or above across Europe will almost triple by 2060 (EUROSTAT 2017).

The macro-level demographics and associated trends mask big differences within ageing populations. There can be as much as 10 years difference in life expectancy within high-income countries, for example in the UK a baby boy born in Kensington and Chelsea has a life expectancy of 83.3 years, compared with a boy born in Glasgow who has a life expectancy 10 years lower (73.0 years) (ONS 2015). For newborn baby girls, life expectancy is highest in Chiltern at 86.7 years and 8 years lower in Glasgow at 78.5 years (ONS 2015; NRS 2016). There is also considerable variation within cities, spatially and socially.

This volume brings together contributions from a broad range of human geographers with different disciplinary perspectives of transport and ageing. This chapter outlines some of the key contemporary issues for an ageing society in terms of transport and mobility, highlights the importance of considering transport and mobility for ageing populations and outlines the contribution that a geographical approach can offer to studies of transport and ageing.

Older People and (Hyper) Mobility

In many ways, we are living in a hypermobile society. Humans have always been mobile, but the intensity and scale of contemporary mobility (Kwan and Schwanen 2016) are greater than in the past. We are traversing greater distances to reach destinations for work, shopping, to access services and health care and for recreation and simply to stay connected to people. We live further away from work and are more mobile than previous generations in terms of moving between jobs. We end up with many more connections to dispersed communities we wish to stay in contact with. Older populations today are more mobile for longer periods of time and in many cases have a high degree of leisure mobility; for example, Andrews et al. (2007) discuss commodification of active ageing and mobile leisure practices for active retirees.

At the same time, due to the balance shifting from infectious diseases to chronic health conditions, there is suggestion that proportion of years of life spent in good health is falling (ONS 2017). Long-term health issues can reduce an individual's ability to be mobile. Despite increases in mobility, in many countries, it is the oldest age group that face the biggest barriers to getting out and about. Older people can therefore face mobility deprivation, feel disconnected to society and be unable to do the things they want to simply because they are unable to get to these things. This disconnection can have wide-ranging physical and mental health implications. Reduced mobility can lead to other health problems such as obesity, heart conditions and increased falls risk.

Traversing large distances can be more difficult for us as we age. Physiology varies hugely between people, but declining eyesight, hearing, muscle strength and cognition can make mobility harder to achieve for many older people. Older people are much more likely than other age groups to spend more time close to home and in the local neighbourhood, especially after retiring from work (Baltes et al. 1999). In a hypermobile society, this home and neighbourhood area can feel quiet and neglected, as other age groups are out travelling to and from work and leisure, rather than engaging with the local area. Suburban neighbourhoods can be quiet places devoid of services and shops, a place that can feel quite empty for an older person. As in high income counties, city centres in low- to middle-income countries are gradually becoming the preserve of big business and the wealthy with ordinary families and older people moving away to the edges of city centres, dispersing social networks and connections and increasing distances needed to access services and shops. People who live in suburban areas, a decision which may have made sense earlier in life given they can offer more space for bringing up a family, tend to find themselves isolated in these 'commuter communities', with housing mismatched to needs as they age and often with little desire or financial resources to move (Howe 2013).

Hence, motorised transport is more important to older people than ever before. The car has become central to this hyperconnectivity, affording the traversing of long distances without recourse to large physical exertion. In high-income and a growing number of low- to middle-income countries, societies and built environments have become so organised around the car, that those without a vehicle can become socially excluded. A divide occurs between those who can benefit from private vehicle ownership and those that experience the wider negative externalities of the car and car-dependent societies, including pollution, severance of communities, crashes and associated casualties. As older people have to give up driving licences, they become at risk of exclusion in car-centric societies.

Gattrell (2013) contends that mobilities scholars have not made much connection between mobility and well-being. Yet, when thinking about transport and ageing, the links between mobility and well-being are quite pronounced. Being mobile in later life is linked to quality of life (Schlag et al. 1996). Giving up driving, in particular, is linked to decrease in well-being and an increase in mental and physical health problems. This is due both to a reduction in ability to get out and about but also related to psychological issues associated with freedom, status, norms and independence (Edwards et al. 2009; Fonda et al. 2001; Ling and Mannion 1995; Marottoli et al. 1997, 2000; Mezuk and Rebok 2008; Musselwhite and Haddad 2010; Musselwhite and Shergold 2013; Peel et al. 2002; Ragland et al. 2005; Windsor et al. 2007; Ziegler and Schwanen 2011).

In higher-income countries, the use of the car has become ubiquitous across the lifecourse, resulting in a large increase in both the number of older drivers and the distances travelled. Compared with previous generations, older adults are much more likely to own a vehicle, particularly women (see Chapter 2). At the same time, walking, cycling and non-urban bus use have generally been in decline across age groups, although there are some signs that this is changing. The decline of public transport as a result of car-centric mobility corresponds in many cases with the geographical location of populations of older people, meaning that issues of transport-related exclusion can be particularly significant for older adults in more suburban or rural areas without adequate public transport.

From the point of view of reducing isolation and loneliness and providing safe mobility options for older adults, it is easy to see maintaining driving and private car use for as long as possible as a panacea. From this perspective, emerging transport options such as autonomous vehicles are seen as being a perfect technological solution to mobility problems for older people, affording them high levels of mobility with minimum effort. How far this will be a reality is questionable. Such a perspective is at odds with policy agendas of sustainability and active ageing. The topic of autonomous vehicles is picked up in Chapters 3, 9, and 10. Focussing more on the structural issues of car dependence could mean that alternative modes of transport would provide similar levels of mobility. Transport infrastructure and supporting policies and practices should be designed with an ageing population in mind. In many countries, public transport use, especially bus use, increases in later life (Musselwhite 2017a). However, a major barrier for older people using public transport is feeling unsafe, especially at night (Gilhooly et al. 2002). Accessibility issues are also an issue with step-free access and availability of seats high priorities for older people. Empathetic and friendly staff are vital; a bus driver driving off before the older person has sat down is a major concern, and the presence of a friendly helpful, understanding staff can be an enabler for older people to use the service (Broome et al. 2010). Cycling is a mode that is quite often associated with younger people, yet it can provide a perfect connection between remaining physically active and traversing longer distance, particularly through increasing use of e-bikes for example. Improving cycle accessibility for older people needs further consideration as emphasised in Chapter 6. Increasing policy attention is being paid to the idea of 'active ageing' and the promotion of walking for health among older adults. However, the importance of individuals' interactions with the built environment is important and reveals that walking as a transport practice is about more than travel from A to B, but hangs deeply upon the meanings and experiences in the environment (see Chapter 8).

Traditionally, transport planning has placed emphasis on the reduction of travel time, and transport policy across many countries is intrinsically economic. As a result, transport policies often centre on commuting and travel for work, which results in transport policy centred on inter-urban transport at peak hours. Policies focussed on reducing the time taken, relieving congestion and making transport more efficient can lead to the hypermobility and urban sprawl we have discussed above and not relieve the problems. It is debatable whether this is of benefit to individuals across different contexts and societies. This is especially true in later life, where working is more likely to have ceased or be moved to part-time, where more local rather than inter-urban journeys take place and where a variety of modes might be used. There is much research to say that mobility and transport use can change and alter dramatically at key transition points (Avineri and Goodwin 2010), including those more likely to be faced in later life, such as retirement from full-time employment and the onset of acute or chronic conditions. To make transport policy more relevant to an ageing population, we need to understand mobility from a lifecourse perspective. There is a need to join up health care and social policy with transport policy, to help meet the accessibility demands of an ageing society and not just kowtow to economic notions. Understanding how older people travel during the day is also an issue as older people can be particularly limited in when they travel and where at certain times; fear of crime and safety concerns, problems with seeing in certain types of light and increased fatigue can all reduce the times of day an older person might go out. Vallée (2017) has recently termed this the daycourse of place—thinking about how places may be more or less accessible at different times of day, and this can be the case for older people who often strategise about when to visit certain places based on safety, light or busyness for example.

In an ageing society, it is therefore important to consider what matters beyond the journey time. It has been found, for example, that while journeys may take longer as people age, satisfaction with the length of journey does not (Curl 2013). This is not to say that older people should adapt and cope with adverse circumstances such as taking longer, but that other factors are more important in how older people perceive and experience accessibility. Issues of urban design and provision of toilets or seating may become more important for example. Focussing on issues of access to important services such as health care facilities and open spaces as well as social connectivity are more important than increasing or maintaining mobility in its own right.

The role of the built environment in influencing physical and mental health is well established (Grant et al. 2017). In particular for older adults, well-designed, inclusive and pleasant urban environments containing greenspace can promote physical activity, supporting policy agendas around active ageing. It is important also to consider the way in which people interact with their environment to establish whether environments are supportive, or not, of active mobility for older adults (Curl et al. 2016; and Chapter 6). While the built environment plays a critical role, assuming a deterministic relationship is problematic (Andrews et al. 2012).

Mobility, Affect and Aesthetics

We tend to think of satisfying accessibility needs through corporeal or literal mobility (Parkhurst et al. 2014). That is that mobility requires physical movement. Often policy and practice identifies solutions for maintaining levels of literal mobility for older people, who may be experiencing declining physical mobility and may have given up driving. The deficit in literal mobility is often seen as problematic in later life, and ways of improving literal mobility are identified (see Musselwhite and Haddad 2017, for examples and reviews). Technological moves towards partly or fully automated vehicles may support drivers who struggle with driver tasks, lengthening the time that people can rely on car-based mobility for example.

Mobility also has social or affective dimensions. Mobility is not always just a means to an end, it can in itself be linked to an individual's quality of life. For example, Clayton and Musselwhite (2013) found the kinaesthetic pleasure of mobility as experienced while cycling is in itself a motivation for using that mode. The bus can provide a third space, for social interaction, people watching or simply watching the world go by (Andrews 2012; Musselwhite 2017a). There is a feeling of satisfaction for completing a long or difficult drive among some older people that is missed when they use other passive transport options (Musselwhite and Haddad 2010, 2017). In addition, then journey itself can be rewarding, travelling past green (e.g. nature, trees, fields, forests, etc.) or blue space (e.g. seascapes, rivers, lakes, etc.) and even seeing familiar sights signifying home or place (Musselwhite 2017b). The car can meet these affective needs easily, someone can simply 'go out for a drive' or 'take the long way home' to see a particular feature of the environment, and these elements are really missed when someone gives up driving (Musselwhite 2017b; Musselwhite and Haddad 2017). Gattrell (2013) has discussed the idea of therapeutic mobilities as an extension to the literature on therapeutic environments.

Conceptualising Mobility and Well-being in Later Life

Musselwhite and Haddad (2010) explain mobility for older people in terms of individual need, highlighting the importance of three different motivations for mobility in a hierarchical manner (see Fig. 1.1). At the base of the pyramid is mobility for utilitarian purposes—that is to be mobile in order to get from A to B as easily, cheaply and efficiently as possible. Once this can be satisfied, the need to be mobile in terms of affective or emotive motivations is the next level of importance, including how mobility provides a sense of independence, freedom and is related to roles and status. Finally, a top level of need is the motivation to be mobile for aesthetic purposes, related to both intrinsic factors of the journey itself and to the discretionary nature of viewing the outside world. Mollenkopf et al. (2011) addressed affective needs in more detail, explaining the importance of out of home mobility as an emotional experience, to note physical movement as a basic human need, to stress that mobility should be seen an expression of personal autonomy

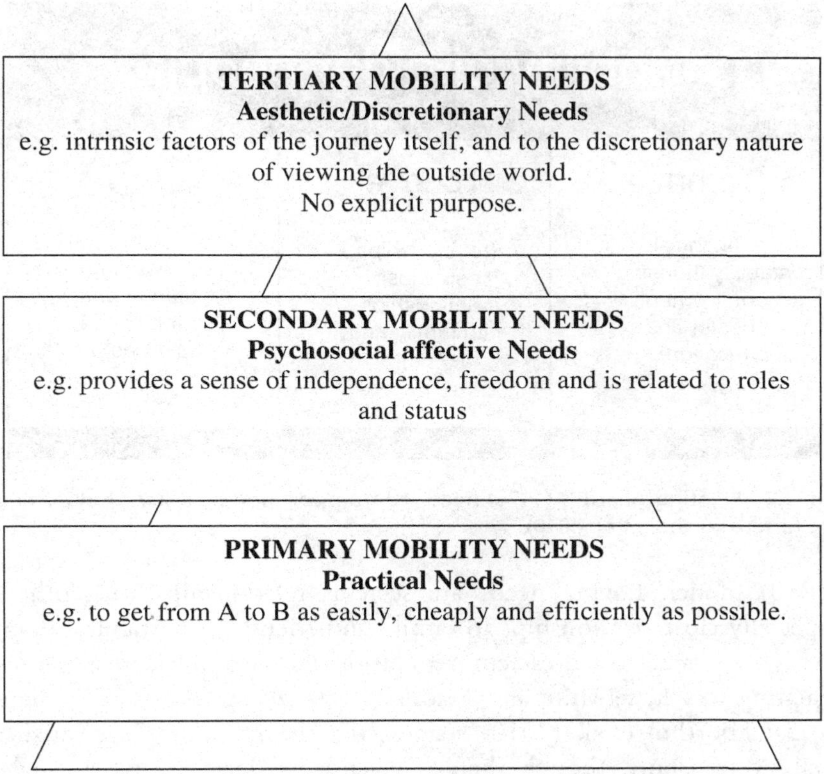

TERTIARY MOBILITY NEEDS
Aesthetic/Discretionary Needs
e.g. intrinsic factors of the journey itself, and to the discretionary nature of viewing the outside world.
No explicit purpose.

SECONDARY MOBILITY NEEDS
Psychosocial affective Needs
e.g. provides a sense of independence, freedom and is related to roles and status

PRIMARY MOBILITY NEEDS
Practical Needs
e.g. to get from A to B as easily, cheaply and efficiently as possible.

Fig. 1.1 Different levels of mobility need among older people (adapted from Musselwhite & Haddad 2010)

and freedom and stimulation. The absence of movement is equated with the end of life, and movement is an expression of the person's life force.

Hjorthol et al. (2010) related Allardt's (1975) model of well-being to transport in later life (see Fig. 1.2). Allardt (1975) suggested three levels of need that have to be satisfied to support well-being: having, loving and being. Having needs relate to financial stability, housing, employment, health and education. In this respect, they are utilitarian in nature, similar to the bottom level of Musselwhite and Haddad's

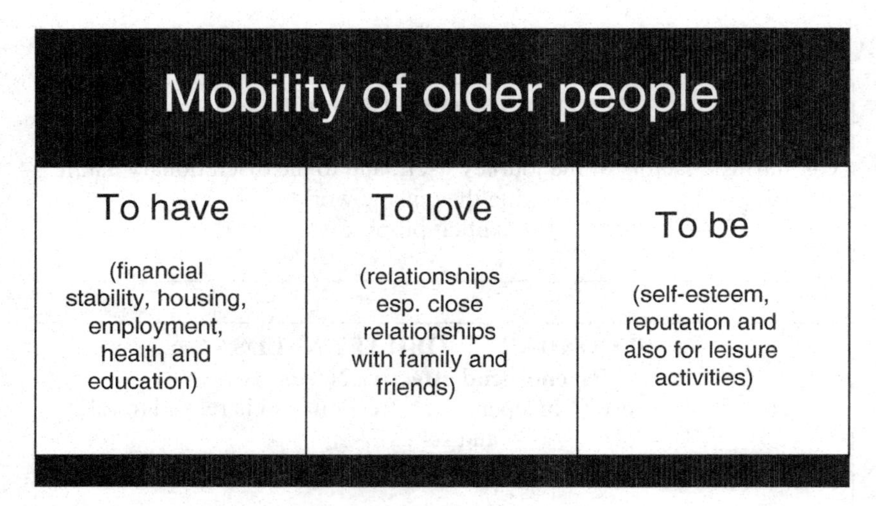

Fig. 1.2 Relating Allardt's (1975) needs to transport and mobility of older people (after Hjorthol et al. 2010)

(2010) model. Loving needs are seen as relationships with others, especially close relationships to family and friends in particular. Being needs are related to self-esteem, reputation and also to leisure activities, equating to Musselwhite and Haddad's (2010) affective and aesthetic needs. Hjorthol et al. (2010) suggest that having needs are satisfied well by transport, though there are gender differences, where males are more satisfied than females, suggesting they often hold more of the transport resource in a household. Loving and being needs are especially in demand but invariably met through current transport systems. Hjorthol et al. (2010) suggest mobility often encompasses more than one and often all three needs. Shopping, for example, covers two elements both having (the need to purchase goods) and loving (the social nature of shopping). In many cases, shopping would often be combined with a social trip, for example, a visit to a café to interact and meet with others.

Needs-based approaches can be criticised for their rather individual and static nature. We don't necessarily always know what we need, and are therefore unable to articulate needs in a study or indeed explicitly be motivated by them on a day-to-day basis.

Needs can also be generated or uncovered as we interact with the environment around us but not recalled in a static interview (see Chapter 8 for the benefits of mobile methods in this respect). This can lead to approaches that bring in wider social processes that may interact with the individual, for example ecological approaches, where the individual is part of a wider social process and interacts with the environment around them. Webber et al. (2010) devised a conical-shaped model based on ascending levels from the individual, their room, then their home outwards through neighbourhoods and neighbouring areas to the world. Each layer has five determinants (cognitive, psychosocial, physical, environmental and financial), with gender, culture and biography (personal life history) viewed as cross-cutting influences. Webber et al. (2010) stress that mobility is literally moving oneself (e.g. by walking, by using assistive devices or by using transportation) within these environments from home, to neighbourhood, and regions beyond. Each of the five determinants affects mobility in different ways, creating barriers and enablers to moving through the different layers. Musselwhite (2016) has devised an age-friendly transport system approach utilising Bronfenbrenner's (1979, 1989, 2005) ecological model, to show how different elements of the environment interact with each other and impact upon the individual and are impacted upon by the individual. Musselwhite's (2016) model starts with the person in the centre, with concentric circles spreading outwards to laws, policy and plans at the outside, connected at the neighbourhood with public and community transport provision between them (see Fig. 1.3). A major issue with ecological approaches is simply how complex they can be, with many interacting layers. Their complexity can make implementing interventions aimed at improving mobility for older people hard to identify where in the structure and in which layer interventions should be targeted. Nevertheless, ecological approaches show how differing factors interact with one another, and it is important to recognise the interactions and impacts of multiscalar factors. Socio-ecological models have also become popular in health research as the broader social determinants of health are recognised from a health promotion perspective.

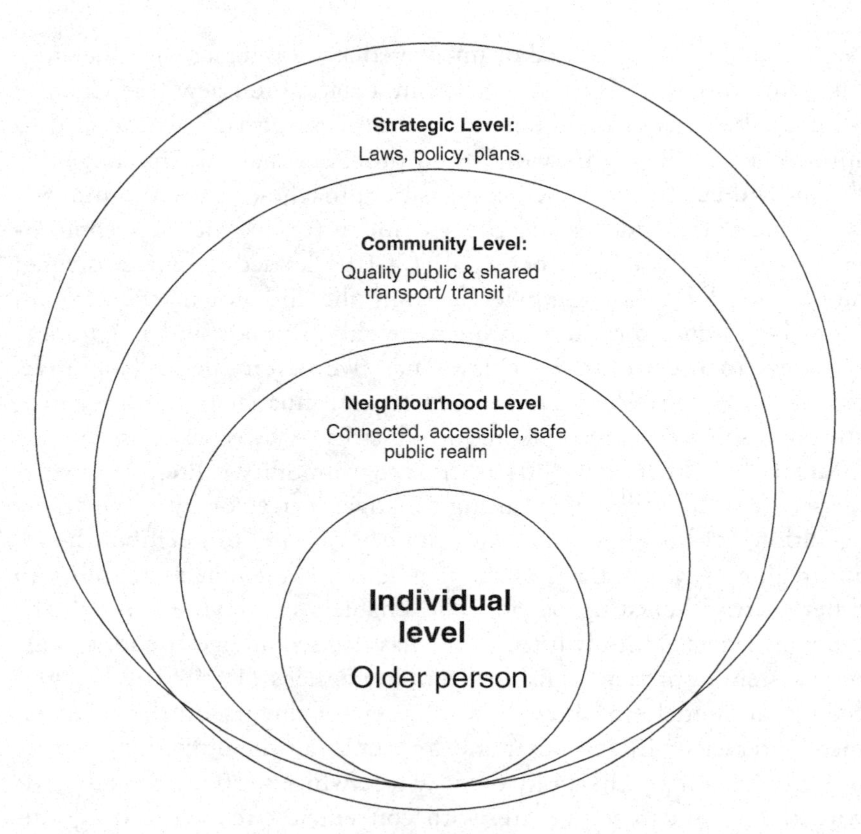

Fig. 1.3 Musselwhite's (2016) ecological model of transport and mobility in later life

Technology, Ageing and Mobility

Technology has a large impact on mobility across the lifecourse and can become particularly pertinent as we age. Technology might be directly transport related, such as real-time travel information or autonomous cars or technologies which indirectly impact on mobility for example changes in shopping behaviours due to increase in use of cars and fridge-freezer technology. Technological changes are almost always discussed in positive terms, through reducing wasted time, physical effort and improved efficiencies. Questionable as that may be, it is even further debatable as to whether technological advances improve mobility and accessibility, especially when considering older populations. For example, although out of town shopping can reduce overheads for retailers, potentially leading to cheaper goods to the customer, the external costs are passed on to society through increasing car-based travel and reducing the amount of shopping in city, town or village centres. Older people who may have restricted mobility may suffer from declining local services and shops as they agglomerate on motorway or road corridors inaccessible to those without a car themselves, increasing exclusion, loneliness and isolation. Without appropriate governance, electric vehicles and driverless cars are likely to further perpetuate, or at least not restrain hypermobility. Additional information, such as real-time bus information and travel information apps, can be beneficial to older people, though older people are the group still most likely to value the importance of talking to people. They are more likely to trust information if it is given from authority figures, for example bus drivers and railway staff, and want the staff to be friendly and approachable (Musselwhite, 2011, 2017a). Chapter 3 of this volume covers the theme of technologies, mobility and ageing in more detail, and Chapter 9 takes a critical perspective on emerging transport technologies.

Changes in telehealth, telemedicine and mobile e-health mean literal visits to the hospital and doctors can be reduced or in some case eliminated altogether. Access to health care varies considerably across different countries and regions. An example is many countries in Africa, where huge inequalities exist in provision health care especially between rural and urban areas. The dispersed nature of populations and health care in Africa have resulted in the World Health Organisation promoting e-health projects aimed at addressing physical inaccessibility to health care (see Chapter 4 for examples).

Parkhurst et al.'s (2014) linear model of mobility places mobilities aided by technology in relation to literal or corporeal technology. They note literal mobility might only be one of the elements where mobility in later life functions, providing a continuum from linear to imaginative through potential and virtual domains. In a hypermobile world, mobility gets most attention in the literal or corporeal domain, where mobility equates to the moving of individuals across space. However, people are also related to their mobility through potential, virtual and imaginative mobility, and these need attention in the literature. The pervasive nature of the perception that mobility can occur anytime and anywhere is encapsulated by potential mobility (Metz 2000). The car sitting on the drive waiting to go out should the individual require it, can be as important as actually doing the journey itself, termed by Nordbakke and Schwanen (2014) as motility. Literal mobility can be supported by or substituted with virtual mobility or accessibility, the ability to complete shopping online, chat to family and friends, view webcams or to have health appointments without literally having to move large distances, to be virtually present in places and even virtually co-present with others is easily satisfied with technology (Chapter 5). Though the question remains what is missed from not literally being immersed in the place or co-present with others that are not there in a virtual connection? We should also consider what is missed in terms of physical activity, which is important for health and well-being across the life-course, but can become particularly important for older adults. Finally, imaginative mobility encapsulates two different propositions: (1) connection to distant mobility through reminiscing and imagining the places visited and (2) connection to the outdoors through interacting with the world from a distance. Ziegler and Schwanen (2011) provide a similar taxonomy in their study of older people in County Durham, UK. They propose five elements of mobility: (1) mobility practices are literal mobility; (2) mobility of the self, is the disposition to connect to the world; (3) attitudes to mobility and relationship of self to mobility; (4) imaginative mobility is where memory and imagination link mobility and the self through recollection or construction; and (5) electronic mobility, using internet, telephone and television to maintain mobility needs. Mobility practices equate with literal mobility in Parkhurst et al.'s (2014) model, showing the importance of literal and temporal practices to maintain daily life. Mobility of the self is similar to Metz (2000) and Parkhurst et al.'s (2014) potential for travel, but is linked more towards a will to remain connected socially than to a specific form of transport.

Electronic and imaginary mobilities clearly map to virtual and imaginative mobilities proposed by Parkhurst et al. (2014). Virtual and imaginative mobilities are both explored in more detail in Chapter 5.

Geographical Perspectives

Transport geography has been called both peripheral (Hanson 2000) and central (Shaw and Sidaway 2011) to debates in human geography. The chapters in this book demonstrate that transport and mobilities are critical when considering ageing. In calling for human geographers to engage with transport more explicitly, Shaw and Sidaway (2011) argue that much geographical work implicitly relates to transport, but is undertaken by those who would not call themselves transport geographers and as such transport is often considered as a given rather than scrutinised. Given that mobility and transport are central to many of the issues of ageing populations, we would argue that geographical gerontology is one such field where transport is implied and underlies the issues but is not given explicit consideration. For example, Andrews et al. (2009) outline five key areas where geography and gerontology intersect: space and the macroscale; population ageing and movement; services, planning and policy; health and living environment; and place and the micro-scale. Transport intersects all of these domains, yet is not mentioned at all in the review, and mobility is only mentioned in relation to migration. There is no mention of the importance of older adults' daily mobilities, despite this being core to many the areas of study mentioned. Andrews et al. (2007) suggest that geographical gerontology has moved beyond concerns of health to consider social and cultural aspects of older people's lives. In a similar vein, transport geographers' increased engagement with the mobilities literature means that greater attention is given to social and cultural aspects of travel. Given the broader convergence of research around transport and health, a volume dedicated to the geographies of transport and ageing is a timely contribution to contemporary debates in transport geography, geographical gerontology and human geography more broadly. Issues of transport and ageing relate to many other sub-disciplines within human geography, including but not limited to: geographies of health and well-being; emotional geographies; urban geography; rural geography; and memory research. This is not therefore a transport geography or gerontological geography volume, but presents a broad range of human geographical perspectives on transport and ageing, which we believe are topics around which geographers from a range of sub-disciplines can converse and provide critical analysis on transport, new transport technologies and the health and well-being of ageing populations.

Although transport geography has been critiqued for its generalist and positivist approaches, the importance of context, situation and place has grown in recent years (Schwanen 2017). It is this consideration of the social, temporal and spatial variation in transport and ageing which the chapters in this book contribute to. Much of the emergent research on the role of the built environment (including transport) and health and well-being takes a deterministic approach to the cause and effect (Andrews et al. 2012; Davison and Curl 2014; Schwanen 2016). The need to move away from separating contextual and compositional influences on health and recognise diversity of relationships between people and place has been highlighted (Cummins et al. 2007).

We have divided this collection into sections. This chapter along with the following two chapters set the context for what is to follow. Chapter 2 analyses generational changes in car-based mobility, paying attention to gender and income differences. Next, Chapter 3 reviews issues relating to older adults' transport and technology, a theme which is drawn out in many of the following chapters. Although we have divided the following two sections into rural and urban geographies of ageing, it is important to be clear that this is a continuum. The 'rural' chapters focus on empirical work undertaken in rural areas in Tanzania (Chapter 4) and Scotland (Chapter 5), and both consider the role of virtual or imaginative mobilities. Next, three chapters focus on work undertaken in more urban environments. Chapter 6 discuss environmental and social factors influencing cycling; Chapter 7 focuses on the urban built environment and mobility in India, and finally, Chapter 8 discuss approaches to researching urban environments through go-along interviews. In the final section, two chapters explore the future of ageing and transport. First, Chapter 9 takes a scenarios approach to discussing what implications autonomous vehicles might have for ageing populations, and finally, Chapter 10 discuss policy context and policy implications.

Although we have divided the book into these sections, there are some clear themes running throughout the volume relating to technology, health and well-being, virtual mobility, co-production, safety and security, relationality, intergenerational intersectionality, gender, perception and meanings, power and autonomy. These themes highlight the contribution of geographical perspectives to understanding the diverse, heterogeneous and complex issues of transport daily mobility in an ageing society.

Conclusion

Studying older people's transport and mobility using a geographical lens highlights the importance of conceptualising mobility as wider than practical, utilitarian and deterministic approaches. While literal mobility is important, over-concentration on it continues to fuel hypermobility by reinforcing the needs for high levels of mobility to remain connected to society and to the things we want to do as we age. Transport has been described as both liberating and enslaving (Shaw and Sidaway 2011), and we can see this in particular when discussing transport and ageing. While movement and mobility in themselves are important for social connectedness, physical fitness and therefore for physical and mental well-being, hypermobility and the demands of needing to travel further create issues of isolation and exclusion for those who do not have the means or ability to travel. Instead of thinking about how we can continue to be mobile, more consideration needs to be given to thinking about how social, political and physical environments can allow ageing populations to maintain accessibility, while moving less (Shaw and Sidaway 2011).

Older adults' daily mobilties involve affective, emotive and psychosocial components. It sits within wider social contexts, being part of imagined or virtual and technological societies. Understanding mobility suggests mobility itself is important, but sometimes less is more and depth over breadth can be rich and rewarding for older people. Re-imagining mobility in terms of small, local, short movement in the neighbourhood reintroduces a need to look at how technology can support mobility and how we need to focus on walking and cycling to support literal mobility. Planning for mobility for older people needs to take into account an ageing society. To fully embrace mobility in its geographical context then a change of research question and focus is often required, not to understand how we can prolong and maximise distances travelled but to look at improving and enriching local nuanced mobilities, looking to improve accessibility rather than simply adding more mobility.

References

Allardt, E. 1975. Dimensions of Welfare in a Comparative Scandinavian Study. Research Report, 9, Helsinki Research Group for Comparative Sociology, University of Helsinki.

Andrews, G. 2012. Just the Ticket? Understanding the Wide-Ranging Benefits of England's Concessionary Fares Policy. Project Report. London, UK: Age UK.

Andrews, G.J., M. Cutchin, K. McCracken, D.R. Phillips, and J. Wiles. 2007. Geographical Gerontology: The Constitution of a Discipline. *Social Science and Medicine* 65 (1): 151–168.

Andrews, G.J., C.M. Milligan, D.R. Phillips, and M.W. Skinner. 2009. Geographical Gerontology: Mapping a Disciplinary Intersection. *Geography Compass* 3 (5): 1641–1659.

Avineri, E., and P. Goodwin. 2010. *Individual Behaviour Change: Evidence in Transport and Public Health*. Project Report. London: Department for Transport. http://eprints.uwe.ac.uk/11211.

Baltes, M.M., I. Maas, H.-U. Wilms, M.F. Borchelt, and T. Little. 1999. Everyday Competence in Old and Very Old Age: Theoretical Considerations and Empirical Findings. In *The Berlin Aging Study*, ed. P.B. Baltes and K.U. Mayer, 384–402. Cambridge, UK: Cambridge University Press.

Bronfenbrenner, U. 1979. *Ecology of Human Development*. Cambridge, MA: Harvard University Press.

Bronfenbrenner, U. 1989. Ecological Systems Theory. *Annals of Child Development* 6: 185–246.
Bronfenbrenner, U. 2005. *Making Human Beings Human: Bioecological Perspectives on Human Development*. Thousand Oaks, CA: Sage.
Broome, K., E. Nalder, L. Worrall, and D. Boldy. 2010. Age-Friendly Buses? A Comparison of Reported Barriers and Facilitators to Bus Use for Younger and Older Adults. *Australias Journal Ageing* 29 (1): 33–38.
Clayton, W., and C.B.A. Musselwhite. 2013. Exploring Changes to Cycle Infrastructure to Improve the Experience of Cycling for Families. *Journal of Transport Geography* 33: 54–61.
Cummins, S., S. Curtis, A.V. Diez-Roux, and S. Macintyre. 2007. Understanding and Representing 'Place' in Health Research: A Relational Approach. *Social Science and Medicine* 65 (9): 1825–1838.
Curl, A. 2013. Measuring What Matters: Comparing the Lived Experience to Objective Measures of Accessibility. PhD thesis, University of Aberdeen.
Curl, A., C.W. Thompson, P. Aspinall, and M. Ormerod. 2016. Developing an Audit Checklist to Assess Outdoor Falls Risk. *Proceedings of the Institution of Civil Engineers: Urban Design and Planning* 169 (3): 138–153.
Davison, L., and A. Curl. 2014. A Transport and Health Geography Perspective on Walking and Cycling. *Journal of Transport and Health* 1 (4): 341–345.
Edwards, J.D., M. Perkins, L.A. Ross, and S.L. Reynolds. 2009. Driving Status and Threeyear Mortality Among Community-Dwelling Older Adults. *Journal of Gerontology Series A: Biological Sciences and Medical Sciences* 64: 300–305.
EUROSTAT. 2017. Population Structure and Ageing. http://ec.europa.eu/eurostat/statistics-explained/index.php/Population_structure_and_ageing.
Fonda, S.J., R.B. Wallace, and A.R. Herzog. 2001. Changes in Driving Patterns and Worsening Depressive Symptoms Among Older Adults. *The Journal of Gerontology, Series B: Psychological Sciences and Social Sciences* 56 (6): 343–351.
Gatrell, A.C. 2013. Therapeutic Mobilities: Walking and 'Steps' to Wellbeing and Health. *Health and Place* 22: 98–106.
Gilhooly, M.L.M., K. Hamilton, M. O'Neill, J. Gow, N. Webster, F. Pike, and C. Bainbridge. 2002. Transport and Ageing: Extending Quality of Life via Public and Private Transport. ESCR Report L48025025, Brunel University Research Archive.
Grant, M., C. Brown, W.T. Caiaffa, A. Capon, J. Corburn, C. Coutts, et al. 2017. *Cities and Health: An Evolving Global Conversation. Cities and Health*.
Hanson, S. 2000. Transportation: Hooked on Speed, Eyeing Sustainability. In *A Companion to Economic Geography*, ed. E. Sheppard and T.J. Barnes, 468–483. Oxford: Blackwell.

Hjorthol, R.J., L. Levin, and A. Siren. 2010. Mobility in Different Generations of Older Persons: The Development of Daily Travel in Different Cohorts in Denmark, Norway and Sweden. *Journal of Transport Geography* 18 (5): 624–633.

Howe, D. 2013. Planning for Aging Involves Planning for Life (Ch 13). In *Policy, Planning and People*, ed. N. Carmon and S. Fainstein. Philadelphia: University of Pennsylvania Press.

Kwan, M.P., and T. Schwanen. 2016. Geographies of Mobility. *Annals of the American Association of Geographers* 102 (2): 243–256.

Ling, D.J., and R. Mannion. 1995. Enhanced Mobility and Quality of Life of Older People: Assessment of Economic and Social Benefits of Dial-a-Ride Services. In *Proceedings of the Seventh International Conference on Transport and Mobility for Older and Disabled People*, vol. 1. London: DETR.

Marottoli, R.A., C.F. Mendes de Leon, T.A. Glass, C.S. Williams, L.M. Cooney Jr., L.F. Berkman, and M.E. Tinetti. 1997. Driving Cessation and Increased Depressive Symptoms: Prospective Evidence from the New Haven EPESE. *Journal of the American Geriatric Society* 45 (2): 202–206.

Marottoli, R.A., C.F. Mendes de Leon, T.A. Glass, C.S. Williams, L.M. Cooney, and L.F. Berkman. 2000. Consequences of Driving Cessation: Decreased Out-of-Home Activity Levels. *Journals of Gerontology. Series B, Psychological Sciences and Social Sciences* 55B (6): 334–340.

Metz, D. 2000. Mobility of Older People and Their Quality of Life. *Transport Policy* 7: 149–152.

Mezuk, B., and G.W. Rebok. 2008. Social Integration and Social Support Among Older Adults Following Driving Cessation. *Journal of Gerontology Social Science* 63B: 298–303.

Mollenkopf, H., A. Hieber, and H.-W. Wahl. 2011. Continuity and Change in Older Adults' Perceptions of Ou-of-Home Mobility Over Ten Years: A Qualitative-Quantitative Approach. *Ageing & Society* 31: 782–802.

Musselwhite, C. 2011. Successfully Giving Up Driving for Older People. Discussion Paper. International Longevity Centre—UK.

Musselwhite, C.B.A. 2016. Vision for an Age Friendly Transport System in Wales. *EnvisAGE, Age Cymru* 11: 14–23.

Musselwhite, C.B.A. 2017a. Public and Community Transport. In *Transport, Travel and Later Life*, ed. Charles Musselwhite, vol. 10, 117–128. Transport and Sustainability. Bradford, UK: Emerald Publishing Limited.

Musselwhite, C.B.A. 2017b. Exploring the Importance of Discretionary Mobility in Later Life. *Working with Older People* 21 (1): 49–58.

Musselwhite, C., and H. Haddad. 2010. Mobility, Accessibility and Quality of Later Life. *Quality in Ageing and Older Adults* 11 (1): 25–37.

Musselwhite, C.B.A., and H. Haddad. 2017. The Travel Needs of Older People and What Happens When People Give-Up Driving. In *Transport, Travel and Later Life*, ed. C. Musselwhite, vol. 10, 93–115. Transport and Sustainability Series. Bingley, UK: Emerald Publishing Limited.

Musselwhite, C.B.A., and I. Shergold. 2013. Examining the Process of Driving Cessation in Later Life. *European Journal of Ageing* 10 (2): 89–100.

National Records of Scotland (NRS). 2016. Life Expectancy for Administrative Areas Within Scotland, 2013–2015. https://www.nrscotland.gov.uk/statistics-and-data/statistics/statistics-by-theme/life-expectancy/life-expectancy-in-scottish-areas/2013-2015.

Nordbakke, S., and T. Schwanen. 2014. Wellbeing and Mobility: A Theoretical Framework and Literature Review Focusing on Older People. *Mobilities* 9 (1): 104–129.

ONS. 2015. Life Expectancy at Birth and at Age 65 by Local Areas in England and Wales: 2012–2014. Titchfield, UK: Office for National Statistics. https://www.ons.gov.uk/peoplepopulationandcommunity/birthsdeathsandmarriages/lifeexpectancies/bulletins/lifeexpectancyatbirthandatage65bylocalareasinenglandandwales/2015-11-04.

ONS. 2017. Health State Life Expectancies, UK: 2013–2015 (with April, 2017 correction). Titchfield, UK: Office for National Statistics. https://www.ons.gov.uk/peoplepopulationandcommunity/healthandsocialcare/healthandlifeexpectancies/bulletins/healthstatelifeexpectanciesuk/2013to2015/previous/v2.

Parkhurst, G., K. Galvin, C. Musselwhite, J. Phillips, I. Shergold, and L. Todres. 2014. Beyond Transport: Understanding the Role of Mobilities in Connecting Rural Elders in Civic Society. In *Countryside Connections: Older People, Community and Place in Rural Britain*, ed. C. Hennesey, R. Means, and V. Burholt, 125–175. Bristol: Policy Press.

Peel, N., J. Westmoreland, and M. Steinberg. 2002. Transport Safety for Older People: A Study of Their Experiences, Perceptions and Management Needs. *Injury Control & Safety Promotion* 9: 19–24.

Ragland, D.R., W.A. Satariano, and K.E. MacLeod. 2005. Driving Cessation and Increased Depressive Symptoms. *The Journals of Gerontology Series A: Biological Sciences and Medical Sciences* 60: 399–403.

Schlag, B., U. Schwenkhagen, and U. Trankle. 1996. Transportation for the Elderly: Towards a User-Friendly Combination of Private and Public Transport. *IATSS Research* 20 (1): 75–82.

Schwanen, T. 2016. Geographies of Transport I: Reinventing a Field? *Progress in Human Geography* 40 (1): 126–137.

Schwanen, T. 2017. Geographies of Transport II: Reconciling the General and the Particular. *Progress in Human Geography* 41 (3): 355–364.

Shaw, J., and J.D. Sidaway. 2011. Making Links: On (Re)engaging with Transport and Transport Geography. *Progress in Human Geography* 35 (4): 502–520.

Valée, J. 2017. The Daycourse of Place. *Social Science and Medicine* 194: 177–181.

Webber, S.C., M.M. Porter, and V.H. Menec. 2010. Mobility in Older Adults: A Comprehensive Framework. *The Gerontologist* 50: 443–450.

WHO. 2015. World Report on Ageing and Health. Geneva, Switzerland: World Health Organisation. http://apps.who.int/iris/bitstream/10665/186463/1/9789240694811_eng.pdf?ua=1.

Windsor, T.D., K.J. Anstey, P. Butterworth, M.A. Luszcz, and G.R. Andrews. 2007. The Role of Perceived Control in Explaining Depressive Symptoms Associated with Driving Cessation in a Longitudinal Study. *The Gerontologist* 47: 215–223.

Ziegler, F., and T. Schwanen. 2011. I Like to Go Out to Be Energised by Different People: An Exploratory Analysis of Mobility and Wellbeing in Later Life. *Ageing & Society* 31 (5): 758–781.

2

Driving Segregation: Age, Gender and Emerging Inequalities

Jon Minton and Julie Clark

Introduction

This chapter highlights the long-term impacts that a deeply embedded preference for car travel present in many high-income countries may have for an ageing society. Automobility has long been the dominant transport system in affluent countries, to such an extent that the 'car system', or 'automobile society', is considered highly path-dependent and self-reproducing (Beckmann 2001; Urry 2004). Preference for travel by car, like many other preferences, is culturally transmitted (Baslington 2008, 2009), and car use has been likened to an addictive habit, much like

J. Minton (✉)
Urban Studies, University of Glasgow, Glasgow, UK
e-mail: Jonathan.Minton@glasgow.ac.uk

J. Clark
Sociology and Social Policy, School of Media, Culture & Society,
University of the West of Scotland, Paisley, UK
e-mail: julie.clark@uws.ac.uk

© The Author(s) 2018
A. Curl and C. Musselwhite (eds.), *Geographies of Transport and Ageing*,
https://doi.org/10.1007/978-3-319-76360-6_2

tobacco use, not least because car use can also induce dependency and damage to health of both users and those nearby (Douglas et al. 2011).

The effective coordination of transport and land use is necessary to prevent social exclusion and ensure that everyone can participate fully in society, having access to social or work opportunities, health services and other amenities (Lucas et al. 2015). However, in a society where urban planning decisions are often made on the assumption of ready access to private motorised transport, two important issues may be overlooked. Firstly, public transport may offer an alternative in name only; cost, physical accessibility, issues around safety, the provision of services and journey times can make public transport an unrealistic 'choice' (Clark and Curl 2016). Secondly, at any given time, driving is not an option for a proportion of the population, whether by reason of age, physical capacity, cost or other constraints.

There are complex but compelling links between life stage and car use, as well as with household wealth, education and other demographic factors, including gender. Evidence from the USA shows that type of residence and the life stage of household members influence how far and how frequently people tend to travel (Lin and Long 2008). Car dependency tends to increase in households after the birth of a child (Lanzendorf 2010). Residential location also affects car ownership and use, with many suburbs effectively only accessible by car. Important life events, including residential relocation due to moving for university and employment or downsizing towards the end of working life, affect both mobility and automobility. Residential shifts also change social networks, potentially influencing the cultural norms and expectations to which people are exposed, including attitudes towards car use, compared with other travel modes (Hopkins and Stephenson 2014; Sharmeen et al. 2014). In more compact, dense urban forms that reduce the physical distance between trip destinations, greater access to multi-modal transport options can reduce levels of car ownership and use (Dieleman et al. 2002; McIntosh et al. 2014).

Considering the physical and social implications of modal choice, car ownership is generally associated with positive health and well-being outcomes. In the case of older people, car ownership is associated with higher quality of life (Knesebeck et al. 2007); conversely, lack of car ownership is associated with poorer general health and, in particular, with increased prevalence of depression and anxiety, obesity and hypertension, and generally less successful ageing—shorter, unhappier lives lived in poorer health (Ellaway et al. 2016; Groffen et al. 2013; Tyrovolas et al. 2017). These associations are indicative of the extent to which, for those who could afford a vehicle, the mobility and autonomy offered by private motorised travel established the car as the mode of choice during the latter half of the twentieth century. Nevertheless, despite what seems a dominant cultural preference for car use, other modes have much to offer. People who use trains or engage in active travel to commute can have higher levels of life satisfaction than car drivers (St-Louis et al. 2014; Thomas and Walker 2015). Furthermore, switching from car use to public transport and active travel is associated with lower body weight, while switching to car use is associated with higher body weight, in almost identical magnitude (Flint et al. 2014; Martin et al. 2015).

From the mid-1990s, there has since been a reversal in what, at times, seemed an unstoppable trend towards more widespread car ownership and use in relatively affluent nations, evidenced by changes in the proportion of people who learn to drive ('licence holders') and, of those who learn, the proportion who own or have regular access to a vehicle ('car access'). Falling levels of licence holding have been identified in at least nine countries in the developed world and, in many countries, high levels of car use amongst older drivers have been partially or wholly offset by reduced levels of car use amongst young adults (Delbosc and Currie 2013; Jones 2014; Kuhnimhof et al. 2013; Lee-Gosselin 2017; Stokes 2013).

At cohort level, the most stark fall in the proportion of drivers has been observed in the 'millennial' generation: people born after 1980 who, in terms of their automobility, have been referred to in the USA as the 'go-nowhere' generation (McDonald 2015). One explanation for this trend might be that millennials defer important life events that lead to increased car use—such as finishing full-time education, marrying or having children—but that, once such events occur, their levels of automobility will increase (Garikapati et al. 2016). However, falling levels of driving in the 2000s have not been accompanied by commensurate increases in the use of alternative travel modes, pointing towards reduced mobility overall, rather than modal shift (Manville et al. 2017). Furthermore, rates of licence holding and car access have not fallen equally for both genders. After many decades in which men tended to drive more than women, in recent years, young adult females now have greater weekly automobility than young adult males of the same age (Tilley and Houston 2016). Amongst young adults within Great Britain, increasing levels of female car use have partially offset falling levels of car use amongst males (Le Vine et al. 2013). Nevertheless, despite these long-term trends, daily mobility remains dominated by the car, with 67% of commuters in England and Wales using cars or vans as their main commute mode in 2011 (Goodman 2013).

Given the significance of driving for achieving full social participation, the aim of this chapter is to graphically explore the complex roles and interactions between generational membership, age, gender and education in understanding both rates of learning to drive and car access amongst those with driving licences. We use a large-scale longitudinal household panel survey based in the UK for our data, and Lexis surfaces—a method often used in demography but rarely for survey data—for our visualisations. Our use of this data and method is exploratory rather than confirmatory, in that it does not involve formal statistical hypothesis testing. Instead, we aim to become more familiar with a large quantity of data, to note and discuss patterns within these data, and to consider the broader sociological, economic and epidemiological implications of these patterns for advancing our understanding of how car society is changing, and the important ways in which inequalities in automobility are either narrowing or expanding.

Investigating Long-Term Trends

Panel data, where the same people and households are repeatedly interviewed over time, are especially useful for understanding long-term social changes (see Baltagi 2005). This research uses the British Household Panel Survey (BHPS), a large longitudinal panel survey first conducted in 1991, and then in every subsequent year until

2009. Although these data are specific to Great Britain, the findings may be of interest and relevance to other countries experiencing similar trends. In the first year, known as a 'wave', of the survey, a representative series of over 5000 households was selected for interview from within 250 locations drawn from the postcode address file, producing a total sample size of over 10,000 individual respondents. In subsequent waves, attempts were made to interview all adult members of the initially selected household (age 16+); if original sample members moved to form new households, then attempts were made to interview them and all members of their new household. However, there was a substantial level of attrition between BHPS waves, as might be expected given the logistical challenges involved in following up so many households and individuals. Additional households were sampled from Wales and Scotland from 1999 and Northern Ireland from 2001. Because a different set of questions was used to elicit information from individuals about licence holding and car access in the first two waves compared with all subsequent waves, all results and analyses presented begin with the third wave (1993). The questions used are discussed below.

Licence Holding, Car Access and Education

Even without car ownership, a driving licence increases motility, insofar as it offers the potential to access private motorised transport (Kaufmann 2002). The licence holder may, financial resources permitting, hire or borrow a car, participate in car share schemes or take up employment that might offer vehicle access. The BHPS variables DRIVER and CARUSE were used to establish firstly whether an individual possessed a driving licence and, subsequently, whether they had access to a car. Adults are first asked, 'Do you have a full driving licence' (the DRIVER variable), and then, if they respond positively to the DRIVER, asked 'Do you normally have access to a car or van whenever you want to use it' (the CARUSE variable).

Added to this information, we used data on highest educational qualification. This can be considered as a proxy for social class, in that it is as an indicator of the relative affluence and advantage of the households that people belonged to as children, and the earnings potential of the households that people form, as they enter adulthood and move through their working lives towards retirement (see Serafino and Tonkin 2014). Throughout the period in which the BHPS was run, the International Standard Classification of Education (ISCED) used a seven-tier grouping of educational classifications, with the following designations—0: pre-primary education; 1: primary education or first stage of basic education; 2: lower secondary education or second stage of basic education; 3: upper secondary education; 4: post-secondary non-tertiary education; 5: first stage of tertiary education; 6: second stage of tertiary education.[1] This analysis has used the ISCED to produce a threefold grouping of populations by highest educational qualifications. Being an international classification system, this allowed people who had not received their education within the UK to be included in the analyses. The sevenfold ISCED groups were categorised into the following three groups for the purposes of this analysis: groups 0, 1 and 2 were collapsed into the category 'no further' education ('Low'); groups 3, 4 and 5 were grouped into the 'further vocational' education group ('Med'); and 6 and 7 were grouped into the category 'further non-vocational' ('High').

Data Visualisation

We explore the data graphically, by arranging either rates of licence holding or car access amongst those with driving licences, as a large matrix of shaded cells or tiles in a levelplot. Each row in these matrices represents people at a particular age, and each column a different year, based on BHPS survey wave. This particular arrangement of data is known as a Lexis surface and is used relatively often within demography, and increasingly in public health, but much more rarely in the broader social sciences (Minton et al. 2013; Schöley and Willekens 2017; Vaupel et al. 1997). The shade of each of the cells in the levelplots indicates, for a particular combination of age and year, either the proportions of sample members with a

[1] http://uis.unesco.org/en/isced-mappings. Accessed 18 April 2017.

driving licence, or the proportion of those sample members with driving licences that also have access to a car. Darker shades indicate higher proportions and lighter shades indicate lower proportions. Colour versions of the levelplots are also available online https://github.com/JonMinton/driving_segregation. Lexis surfaces contain a lot of information, which can be uncovered by identifying patterns of shade within the levelplot. To facilitate discussion of these patterns, we include guide versions of the plots, with labels and other annotation; these are referred to throughout the results section.

Limitations of the Research

The UK Household Longitudinal Study (UKHLS) superseded the BHPS in 2008. Unfortunately, many of the questions and classifications of responses are inconsistent between BHPS and UKHLS, meaning it proved problematic to 'extend' the observations shown above beyond 2009 using the UKHLS. The complex sampling and questionnaire design of the BHPS have both advantages and disadvantages, with the main advantage being that individuals can be tracked through time and so the effect of changes in individual circumstances on other outcomes estimated. Additionally, though the BHPS was initially drawn from a representative sample of the UK population, both selective attrition and the booster samples mean it can become somewhat less representative of the UK population over time (Uhrig 2008). Within the analyses presented here, the BHPS is presented 'as is', without attempts to explicitly follow the same individuals over time or to analyse the influence of specific changes in household or individual circumstance on mobility outcomes; however, the BHPS has been used to allow these analyses to be explored in subsequent research.

Results

In this section, we first examine changing patterns of licence holding. We explore how licence holding has changed over time, looking at gender and generation before going on to consider the role of highest educational qualification. After this, we investigate changing patterns of car access, again, looking initially at gender and generation before going on to examine education.

Exploring the Levelplots: Licence Holding and Car Access

Figure 2.1 shows Lexis surface levelplots for the proportion of BHPS sample members who have a driving licence (subfigures 2.1a–c). Figure 2.2 shows the proportion of those BHPS sample members with a driving licence who also have access to a car or van (subfigures 2.2a–c). Subfigures 2.1c and 2.2c show levelplots separately for each gender (females on the left and males on the right), whereas subfigures 2.1b and 2.2d show the levelplots further subdivided by highest educational qualification. Subfigures 2.1c and 2.2c contain a number of simple labels and divisions, indicating different regions within the Lexis surfaces, which will be referred to in this discussion of the results. Within each levelplot, the shade of a cell indicates the proportion, with black cells indicating 100% and white indicating less than 50%.

Licence Holding: Gender and Generation

Figure 2.1a shows the proportion of the adult BHPS sample who report having a driving licence, from 1993 to 2008, and for all ages from 17 to 80 years. It is clear from the difference in the shade of the right sub-panel (males) compared with the left sub-panel (females) that, historically, a larger proportion of adult males tend to have driving licences than females. A subtler pattern in this figure is suggested by noting that in both panels, and in particular for the female panel, the cell shades tend to be darker near the bottom of the panels than at the top, when looking at both panels from the top to around one-third of the way from the bottom. This indicates that, above around the age of thirty years, younger adults tend to be more likely to have a driving licence than older adults. As largely the same panel of individuals are being

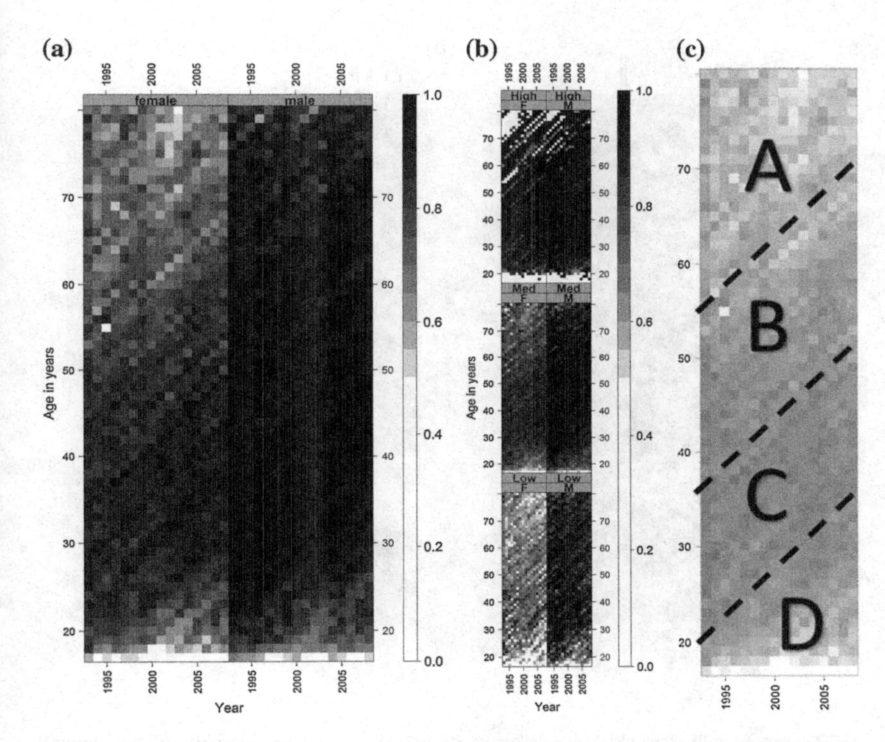

Fig. 2.1 Lexis surfaces showing the proportion of BHPS sample members who have driving licences (a–c). In all figures, year runs horizontally from left to right, and age runs vertically from bottom to top. Within each levelplot, the shade of a cell indicates the proportion, with black indicating 100% and white cells indicating less than 50% or missing values. **(a)** The proportion of BHPS sample respondents holding a driving licence, by gender. **(b)** Licence holding, by gender and highest qualification. **(c)** Licence holding: guide to cohort patterns

followed each year, and only a very small proportion of people possessing driving licences then have these licences revoked and have to take the test again, this difference in shades is suggestive of changes in driving licence ownership rate by cohort, with an overall pattern of successive cohorts being more likely by a given age to possess a driving licence than earlier cohorts at the same age.

By comparing similar regions (combinations of age and year) in the male and female panels, it is also apparent that levels of driving licence

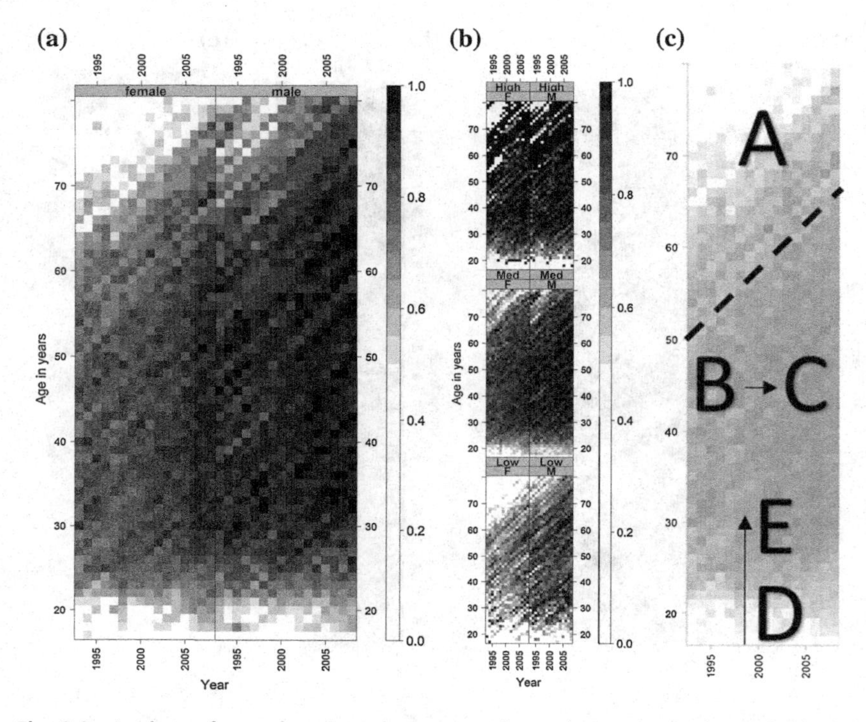

Fig. 2.2 Lexis surfaces showing the proportion of BHPS sample with driving licences, who also have access to a car or van (a–c). (**a**) The proportion of driving licence holders with access to a car or van, by gender. (**b**) The proportion of driving licence holders with access to a car or van, by sex and highest educational qualification. (**c**) Car or van access: guide to cohort patterns

ownership between males and females have tended to converge over successive generations. To look at this further, consider the region indicated by the letter A in Fig. 2.1c, and above the first diagonal dashed line in the figure; this broadly demarks cohorts born before around 1940. Within this broadly defined region of the Lexis surfaces, the cells are much darker for males than for females, with the proportions of males with driving licences around 90%, and the proportions of females with driving licences from similar cohorts ranging from around 55 to 70%. Next, consider the region indicated by B in Fig. 2.1c, demarcated by the first dashed line above and another parallel diagonal dashed line below. This broadly indicates cohorts born between around the early 1940s and the late 1950s. The difference in the cell shade in this region between the male and female panels has reduced, with the proportions of males with driving licences increasing slightly from around 90 to 95% and above, and the corresponding female driving licence ownership rates increasing from around 70% to over 80%. The region indicated by the letter C in the Fig. 2.1c indicates cohorts born from around the early 1960s to around 1975. For these cohorts, the proportion of males with driving licences has remained high, at around 95% or above, whereas the proportion of females with driving licences has increased further, from around 80% to around 90%. Finally, we can consider the bottom-right corners of the panels, indicated by letter D on Fig. 2.1c. This shows driving licence rates for people born after around 1975. What is striking about these younger cohorts is that driving licence rates have fallen for both genders compared with earlier generations, reversing a trend towards higher driving licence ownership, which had been continuing for many generations. It is also noteworthy that these falls in licence rates have been in both genders, reaching around 75–80% for some of the newer cohorts within this Lexis surface region.

Licence Holding: Considering Education

Figure 2.1b shows how rates of driving licence ownership vary by year and age, and further by both gender and highest educational qualification class. The top row shows rates for those with the highest educational grouping ('Further non-vocational', labelled 'High'); the middle for intermediate qualifications ('Further vocational', labelled 'Med') and the bottom row for those with 'no further' education (labelled 'Low'). As before, the left panels show the rates for females and the right panel shows the rates for males. Though each panel is smaller than before, and so it is harder to make out the details of each panel, a number of broad trends and differences between panels are clear, revealing important information about the complex relationship between gender, generation and income.

To learn more about the moderating influence that higher qualifications appear to have on gender differences in automobility, we can compare the overall shade of the left- and right-hand panels in each row.

For the top row, for those whose highest educational qualification is a degree, there is very little difference in shades between these two panels, indicating very little difference in automobility by gender within this high educational subpopulation. The overall shade also tends to be uniformly darker than in any of the other panels, indicating higher driving licence rates overall, which are close to 100% for either gender, at almost all ages, and in almost all years. There are, however, notably more missing values (blank, white cells) for older females than older males, because historically fewer females than males attended university, and so for particular combinations of age, year, gender and educational qualification there were simply no observations in the sample.

By contrast, within the lowest educational qualification group (bottom row panels), there is both a lower proportion of people with driving licences overall than in the other panels, as well as the greatest difference between male and female rates of driving licence ownership. Historically, male levels of driving licence ownership tended to be at around 90%, increasing steadily up to around 95% or above for those cohorts born up to around 1970–1975; by contrast female rates in the earliest cohorts were only slightly above 50%, rising to around 80% by the end of the 'catch up' generation (bottom of region B in Fig. 2.1c).

Within the intermediate qualifications group (middle row panels), there is both an intermediate level of overall disparity in gender mobility (difference in shade between left and right panels) as well as perhaps clearer diagonal 'striation' then in the other panels, suggesting that cohort effects are particularly important in explaining car use in people with intermediate-level qualifications, and that the generational patterns and changes described above for the whole BHPS sample are particularly the cases for the intermediate qualification subpopulation.

Car Access: Gender and Generation

We will now look at trends and patterns in the proportion of the BHPS sample with a driving licence who also state they have access to a car or van. For the whole of the relevant BHPS subsample, this is shown in Fig. 2.2a, with the female panel on the left and the male panel on the right. Figure 2.2c labels some of the regions within the panels, A–E, which will be referred to in the discussion of Figs. 2.2a and c. As with Figs. 2.1a–c, the shade of cells within the panels indicates proportions, with proportions below 0.5 represented by white cells, and higher proportions ranging from 0.5 to 1.0 by successively darker shades.

We will begin by considering the region A, representing those (now post-retirement) cohorts who were around 50 years old or older in 1993, and therefore cohorts born either before or during the Second World War; the very earliest cohorts visible in region A are persons aged 80 years in 1993, and so region A includes some cohorts born from the 1910s to the 1940s. Region A in Figs. 2.2a–c therefore covers a similar range of cohorts to region A Figs. 2.1a–c, and a somewhat similar pattern of change is seen. For women and within region A, there is evidence of successively higher proportions of those with driving licences also having access to a car or van, but with higher proportions of males than females of the same age and in the same cohort. For cohorts born in the 1910s, around 20% of women with driving licences, and around 50% of men with driving licences, had access to a car or van. For cohorts born in the 1920s, the proportion of licensed females with a car or van access rose from slightly under 30% to around 60%; the corresponding change for men in these cohorts was between around 55% and about 80%. For cohorts born in the 1930s, the proportion of licensed females with car or van increased to around 80%, whereas for licensed males it increased to around 90%. Within region A, therefore, the proportion of female drivers with a car or van access increased from around 20 to 80%, and for males from around 50 to 90%.

It is important to note that, within the age range 60–80 years, an age effect is not observed, i.e., the proportion of people with licences with car access does not diminish between ages 60 and 70 or 70 and 80. Within the UK, drivers aged 70 or older need to renew their driving licences every three years, as well as to state if they have developed any medical conditions which may affect their driving, but are not required to retake a driving test.[2] Even though rates of impairment that

[2]https://www.gov.uk/renew-driving-licence-at-70.

may affect driving can be expected to increase with old age, there is no indication, at least up to age 80 years, that this substantively affects this measure of car access, though older car users may drive less often and for smaller distances.

We now consider changes in the proportion of drivers with car access aged between around 30 and 55 years of age, how this proportion has changed from the early 1990s to 2008, and how this change has differed from males and females. This particular pattern of change is represented by the letters B and C in Fig. 2.2c, along with the arrow going left to right. We can see a notable increase in the proportion of licensed women in this age bracket with a car or van access, from around 80% to around 90%. Most of this increase appears to occur fairly suddenly, around 2001–2002, rather than being a gradual change. For men in the same age bracket, there is no equivalent change, with rates between remaining around 90–95% throughout the period of observation.

Finally, we can look at how the proportion of people with licences with access to a car or van changes with age from around the age of 17 to 30 years, as indicated by the letters D and E in Fig. 2.2f, and the vertical arrow pointing upwards. We can see that, within this age range, there is very little difference between genders, and instead age effects dominate. At around 20 years of age, around 55% of those with driving licences also have access to a car or van. By the age of 25, this has increased to around 75–80%, and by the age of 30 to around 90–95% for men throughout the period 1993–2008; for women, rates increased to around 80% by age 30 up to around 2001, and to around 90% from around 2004 to 2008, due to the period-driven change represented by the vector from B to C.

Car Access: Considering Education

Figure 2.2b shows how the proportion of registered drivers with access to a car or van varies by highest educational qualification as well as by gender, age and year. As with Fig. 2.1b, Fig. 2.2b allows the mediating and moderating role of educational qualification on automobility patterns to be better understood. As with Fig. 2.1b, we can see, by comparing the shade of cells in the top-left with the top-right panels that gender differences in this automobility outcome (*realised* automobility) are very low amongst those with a degree or higher qualification, with car or van access levels typically above 90% at all ages above around 30 years for both genders. There is also no apparent historic cohort pattern (region A in Fig. 2.2f) whereby realised automobility increases over successive cohorts born from the 1910s to the 1940s; instead, rates of car or van ownership amongst both males and females with degrees from these cohorts tended to be close to 100%, though with fewer observations for females, leading to a larger number of missing cells. If anything, amongst those with degrees and driving licences, rates of a car or van ownership decreased slightly for cohorts born after Second World War, from around 95–100% for pre-war cohorts, to around 90–95% for cohorts born after 1945.

Both the subpopulations with lower and intermediate educational qualifications (middle panel and low panel) differ from the 'high' qualification group in a number of ways. Firstly, the progressive increases in realised automobility in the oldest cohorts with each successive cohort (region A, covering cohorts born from the 1910s to the mid-1940s), which are seen in the population overall, are clearly evident through the diagonal striation within this region. Comparing equivalent cohorts (e.g., looking at the very top-left corners of each panel to compare the 1915 cohorts), we can see both that realised automobility tended to be lower for females than for males, and for the low education compared with intermediate educational qualification group. For example, rates of realised automobility amongst the oldest cohorts were around 20% for females with 'low' qualification, around 50% for males with 'low' qualifications, around 40% for females with intermediate qualifications, and around 70% for males with intermediate qualifications. For each of these groups, with the exception of females with low qualifications, rates of realised automobility reached around 90% or higher for those born after Second World War (bottom of region A); for females with low qualifications levels reached levels of between 60 and 80% instead.

A second way in which the two bottom rows of panels differ from the top row is that there tends to be more of a difference between females and males in realised automobility rates, with these disparities greatest in the lowest educational group and smaller in the intermediate educational group. A third observation to note is that the rapid rise in automobility rates seen for females overall after around 2001, characterised by the vector B to C in Fig. 2.2c, is clearest to see in the panel for females with intermediate qualifications, though to some extent also evident for females with 'low' qualifications.

Finally, it is important to note that the increasing levels of realised automobility seen between around the age of 20 and 30, as characterised by the vector D to E, are seen for all educational subgroups, and do not appear to differ strongly by gender.

Discussion

The complex patterns of licence holding and car access can, to an extent, be simplified into a series of broad generational 'pen portraits', each differing in terms of automobility and gender equality. From the BHPS sample, it appears that generations born in Great Britain before the Second World War tended to have mixed mobility and high gender inequality, with around 90% of males from this generation possessing a driving licence but only around 70% or so of females likely to have a driving licence; put another way, by the time this generation had reached old age, women were around three or more times more likely not to be able to drive than men of the same age. Any such inequality has important implications for how reliant both older women and older men are on either public transport or friends and relatives with access to a car to travel substantive distances from their homes.

For generations born after the Second World War, and up until the start of the 1960s, there was a catch-up in automobility between the genders, with both genders more likely to own a licence by the time they reached middle age, but with a greater increase in women's automobility than that of men. As the post-war baby boom into adulthood, the expansion in licence holding and car access was met by extensive road-building programmes and, during the latter half of the twentieth century, planning decisions were increasingly made on the assumption that the private car was the mode of the future. It took many years for the phenomenon of induced demand to be recognised: increasing road capacity generates, rather than simply *meets*, demand for car travel (SACTRA 1994). If the generation born from after the Second World War to the start of the 1960s can be characterised by both increasing mobility and increasing equalisation of automobility between genders, the generation born from around 1960 to the mid-1970s may be thought of representing an end point in this journey towards a high mobility and high auto-equality society. Within this high mobility, high auto-equality generation, both males and females were highly likely to possess driving licences, though the proportion of females with licences still remained somewhat lower than for males. This high mobility, high auto-equality generation experienced both the tail end of a decades' long social democratic commitment to high-quality education in primary and secondary school for children, then the transition and embedding of neoliberalism under Thatcher and Major while of working age; put another way, this generation (or at least the start of this generation) both gained from the relatively high tax and high social investment policies of the post-war post-Keynesians while economically dependent children, and to some extent from the low tax and low regulation policies of Thatcherism while income-generating and tax-paying adults. These circumstances favour access to a driving licence and car ownership, as well as an appetite for the road infrastructure and parking required to support driving. Increasing job insecurity, or 'flexibilisation', after Thatcher therefore made traditional single-earner households less economically stable, and so less common, so creating both the opportunity and the necessity for dual-earner households to proliferate. This may have been enabled by, as well as helped consolidate the previous generation's progress towards high female automobility rates, as both the ability to work, as well as to balance work with other commitments, can depend on possessing a driving licence and automobility.

For the generation born after 1975, the trend towards increased automobility seems to have gone into reverse, more quickly than it rose for either gender over previous generations. Although we cannot know the proportions with driving licences in old age, the proportion of both genders with driving licences around the age of 20–25 years is falling and does not appear to be increasing as this generation enters their thirties. Interestingly, though the fall in driving licence rates occurred in both genders, and from levels that were around 5% points higher for males than females, it appears they may be falling to similar levels for both genders, of around 80%. This nascent generation therefore appears from the BHPS data to be characterised by high gender equality but relatively low automobility capability, in the sense of holding a licence.

Examining the experience of different generations with regard to realised automobility—that proportion of licence holders who also have access to a car—there is a similar picture. Contextually, though, the 1950s are considered by contemporary standards to have high levels of structural and cultural inflexibility regarding gender norms and female participation in the workplace, increasing affordability of car ownership at household level, due both to rising household incomes and falling vehicle costs through greater industrialisation, led to both 'two licence' households and then 'two car' households becoming increasingly common. However, although differences between male and female car access are most striking amongst the older generations, they are also more pronounced than the differences between male and female licence holding. This reminds us that, for much of the population, growing equality of licence holding is not necessarily reflected in equality of mobility between genders. Nevertheless, a preference for car access very much dominates these data. Firstly, there is the absence of an age effect for the older generations: there is no indication of a drop off in car access in later years, supporting the idea that people are reluctant to relinquish a car. Secondly, in terms of gender difference, age rather than cohort effects dominate at the younger end of the spectrum, in that there is little difference in car access, reflecting a movement away from conditions of gendered dependence in the fields of mobility and earnings, and towards equality and interdependence.

However, considering the falling levels of car access evident in the younger generation, a synthesis of extant research identified six broad categories of potential explanation for declining car use and its decline: life stage, affordability, location and transport, driving licence regulation, attitudes and e-communication; it found somewhat stronger evidence for life stage explanations (such as having children later, staying with parents and in full-time education longer), and affordability explanations (such as rising costs of insurance, licensing, petrol, vehicles; against falling or stagnant household incomes) than other types, but with no clear single cause (Delbosc and Currie 2013). It may be helpful to think about the various categories of explanation for declining automobility amongst younger cohorts as having either a broadly optimistic choice-based or a broadly pessimistic constraint-based interpretation and implications for future mobilities and inequalities. Our examination of licence holding and car access looking at highest educational qualifications leaves us more inclined towards a constraint-based interpretation, in which automobility has become increasingly unaffordable for substantial sections of society, instead of a predominantly choice-based interpretation in which fewer people either need or want to drive cars.

The fall in licence holding seen for post-1975 cohorts is very clear for people with low or intermediate qualifications but not very pronounced for those with high qualifications. This could partly be because there are fewer observations with which to try to discern this pattern in high educational groups, if they delay learning to drive until after a university degree. A more substantively important implication, however, is that whatever changes have occurred that have led to less automobility overall have had more of an impact on those with fewer qualifications. Furthermore, while gender differences in licence holding have equalised for people with intermediate level qualifications, they have remained or become exacerbated in the case of low-level qualifications, with a fall to lower levels for females with the lowest qualifications compared with males. Similarly, the inequalities evident in patterns of car access, discussed above, are amplified when education is added to the analysis. Considering highest educational qualification as a proxy for social class and relative affluence, the age effect in gaining car access is now no longer strongly differentiated by gender. However, we still see lower levels of car access in the relatively disadvantaged groups with women driving less than men and disparities greatest in the lowest educational group.

Conclusion

In this chapter, we used data visualisation to take a long-term view of mobility and modal choice. In doing so, we consider age in terms of both life stage and cohort, aiming ultimately to highlight some key social challenges posed by a continuing preference for car travel. While our focus here has been on Great Britain, drawing on data from the BHPS, the issues raised will be relevant to other, relatively affluent countries, where planning strategies have favoured personal motorised transport.

It can be tempting, in the digital age, to think of isolation and social exclusion due to poor mobility as being a thing of the past for those in more affluent nations. The visualisations presented here act as a reminder that access to private motorised travel cannot be taken for granted in policy and planning. Older generations, with striking disparities in licence holding and even greater differences in the ability to access a car, are a vital part of our society; moving forward, past the era of the baby boom generation to the millennials, younger people exhibit lower levels of licence holding and car access than in the preceding generation. The data highlight a shift towards greater gender equity in relation to licence holding and car access. However, the analysis, exploring sixteen years of panel data by highest educational qualification, also shows striking disparities in opportunity to access private motorised transport between different social groups. This suggests that getting priced out, rather than *opting* out, of car ownership underpins much of the trend towards reduced mobility. Over the long term, this latter trend has concerning sociological, economic and epidemiological implications. An inclusive society cannot afford to base land use and transport planning decisions on the assumption of car access. As well as directly discriminating against appreciable segments of society, such strategic decisions undermine other modes, either by reducing the market for them or creating an uncongenial physical environment. Without substantial shifts in urban form and investment in public and active travel infrastructure, lack of mobility can further amplify the intergenerational advantage and disadvantage associated with educational qualifications, for the disadvantaged, limiting access to social and economic opportunities as well as spatially sorting people into residential locations where they have access to public transport. On this basis, there is a powerful social justice imperative for multi-modal transport planning, ensuring that 'mobility' does not simply equate to car ownership. In particular, for those in the lowest educational grouping who are least likely to have access to private transport, there is a double disadvantage of more rapid ageing, due to the cumulative effects of social disadvantage on health, compounded by poorer mobility. Considered from a demographic perspective, the lower levels of licence holding and car ownership in the millennial generation offer a window of opportunity to re-evaluate urban policy priorities. Under a *business as usual* scenario, we may see increasing spatial segregation between richer and less mobile, poorer citizens. However, given an appropriate supporting framework, the reduced car use of this cohort could offer a pathway to more active travel and healthier ageing, especially if accessibility planning and technological change make car ownership a less important determinant of mobility as the millennial generation ages. Nevertheless, while a modal mix, including active travel and public transport, might offer satisfaction and opportunities for improved health, land use planning and transport policies must come together to make active travel and public transport viable and desirable choices for all.

We remain hopeful about the potential of technological change to lead to more equitable and effective use of cars, alongside other transport modes, in the coming decades. Car sharing schemes are one of the earliest means by which the car can be used more efficiently as an asset, 'selling mobility instead of cars' (Firnkorn and Müller 2012). The price of ride-sharing could be reduced, and so the attractiveness of ride-sharing more attractive, if the dominant platform operators (such as zimride, blablacar and carpooling.com), who provide the information infrastructure to enable such schemes, were to improve their algorithms to better accommodate 'multi-hop' ride-sharing (Teubner and Flath 2015). Automated vehicles, like car-sharing schemes, offer not just the possibility for replacing or blurring the public/private distinction in car ownership, but for augmenting train and coach use as well (Yap et al. 2016). Autonomous car use and technology is still in its infancy, but offers the potential for cars to become more of an efficiently used shared asset rather than a private asset parked, and so not used, for most of the day (Thomopoulos and Givoni 2015). Much simpler technologies, combined with better infrastructure, offer hope for healthier ageing through active travel. Although even within the famously cycle friendly, Netherlands younger people tend to cycle more than older people, rates of cycling are increasing amongst older adults, including 'baby boomers' found resistant to modal change in the UK; this has partly been attributed to the availability of e-bikes as well as cycle infrastructure, suggesting the potential for technological innovations to increase active travel at older ages (Harms et al. 2014).

References

Baltagi, B.H. 2005. *Econometric Analysis of Panel Data*, 3rd ed. Chichester: Wiley.

Baslington, H. 2008. Travel Socialization: A Social Theory of Travel Mode Behavior. *International Journal of Sustainable Transportation* 2 (2): 91–114. http://www.tandfonline.com/doi/abs/10.1080/15568310601187193.

Baslington, H. 2009. Children's Perceptions of and Attitudes Towards, Transport Modes: Why a Vehicle for Change is Long Overdue. *Children's Geographies* 7 (3): 305–322. http://www.tandfonline.com/doi/full/10.1080/14733280903024472.

Beckmann, J. 2001. Automobility—A Social Problem and Theoretical Concept. *Environment and Planning D: Society and Space* 19 (5): 593–607. http://journals.sagepub.com/doi/10.1068/d222t.

Clark, J., and A. Curl. 2016. Bicycle and Car Share Schemes as Inclusive Modes of Travel? A Socio-Spatial Analysis. *Social Inclusion* 4 (3): 83–99.

Delbosc, A., and G. Currie. 2013. Causes of Youth Licensing Decline: A Synthesis of Evidence. *Transport Reviews* 33 (3): 271–290. http://www.tandfonline.com/doi/abs/10.1080/01441647.2013.801929.

Dieleman, F.M., M. Dijst, and G. Burghouwt. 2002. Urban Form and Travel Behaviour: Micro-Level Household Attributes and Residential Context. *Urban Studies* 39 (3): 507–527. http://journals.sagepub.com/doi/10.1080/00420980220112801.

Douglas, M.J., S.J. Watkins, D.R. Gorman, and M. Higgins. 2011. Are Cars the New Tobacco?. *Journal of Public Health* 33 (2): 160–169. https://academic.oup.com/jpubhealth/article-lookup/doi/10.1093/pubmed/fdr032.

Ellaway, A., L. Macdonald, and A. Kearns. 2016. Are Housing Tenure and Car Access Still Associated with Health? A Repeat Cross-Sectional Study of UK Adults Over a 13-Year Period. *BMJ Open* 6 (11): e012268. http://bmjopen.bmj.com/lookup/doi/10.1136/bmjopen-2016-012268.

Firnkorn, J., and M. Müller. 2012. Selling Mobility Instead of Cars: New Business Strategies of Automakers and the Impact on Private Vehicle Holding. *Business Strategy and the Environment* 21 (4): 264–280. http://doi.wiley.com/10.1002/bse.738.

Flint, E., S. Cummins, and A. Sacker. 2014. Associations Between Active Commuting, Body Fat, and Body Mass Index: Population Based, Cross Sectional Study in the United Kingdom. *BMJ* 349 (Aug 19, 13): g4887. http://www.bmj.com/cgi/doi/10.1136/bmj.g4887.

Garikapati, V.M., R.M. Pendyala, E.A. Morris, P.L. Mokhtarian, and N. McDonald. 2016. Activity Patterns, Time Use, and Travel of Millennials: A Generation in Transition? *Transport Reviews* 36 (5): 558–584. http://www.tandfonline.com/doi/full/10.1080/01441647.2016.1197337.

Goodman, A. 2013. Walking, Cycling and Driving to Work in the English and Welsh 2011 Census: Trends, Socio-Economic Patterning and Relevance to Travel Behaviour in General. *PLoS ONE* 8 (8): e71790. http://dx.plos.org/10.1371/journal.pone.0071790.

Groffen, D.A., A. Koster, H. Bosma, M. van den Akker, T. Aspelund, K.Siggeirsdóttir, G.I. Kempen, J.T. van Eijk, G. Eiriksdottir, P.V. Jónsson, L.J. Launer, V. Gudnason, and T.B. Harris. 2013. Socioeconomic Factors from Midlife Predict Mobility Limitation and Depressed Mood Three Decades Later; Findings from the AGES-Reykjavik Study. *BMC Public Health* 13 (1): 101. http://bmcpublichealth.biomedcentral.com/articles/10.1186/1471-2458-13-101.

Harms, L., L. Bertolini, and M. te Brömmelstroet. 2014. Spatial and Social Variations in Cycling Patterns in a Mature Cycling Country Exploring Differences and Trends. *Journal of Transport & Health* 1 (4): 232–242. http://linkinghub.elsevier.com/retrieve/pii/S2214140514000802.

Hopkins, D., and J. Stephenson. 2014. Generation Y Mobilities Through the Lens of Energy Cultures: A Preliminary Exploration of Mobility Cultures. *Journal of Transport Geography* 38: 88–91.

Jones, P. 2014. The Evolution of Urban Mobility: The Interplay of Academic and Policy Perspectives. *IATSS Research* 38 (1): 7–13. http://linkinghub.elsevier.com/retrieve/pii/S038611121400017X.

Kaufmann, V. 2002. *Re-Thinking Mobility*. Aldershot: Ashgate.

Knesebeck, O. von dem, M. Wahrendorf, M. Hyde, and J. Siegrist. 2007. Socio-Economic Position and Quality of Life Among Older People in 10 European Countries: Results of the SHARE Study. *Ageing and Society* 27 (2): 269–284. http://www.journals.cambridge.org/abstract_S0144686X06005484.

Kuhnimhof, T., D. Zumkeller, and B. Chlond. 2013. Who Made Peak Car, and How? A Breakdown of Trends Over Four Decades in Four Countries. *Transport Reviews* 33 (3): 325–342. http://www.tandfonline.com/doi/abs/10.1080/01441647.2013.801928.

Lanzendorf, M. 2010. Key Events and Their Effect on Mobility Biographies: The Case of Childbirth. *International Journal of Sustainable Transportation* 4 (5): 272–292. http://www.tandfonline.com/doi/abs/10.1080/15568310903145188.

Le Vine, S., P. Jones, and J. Polak. 2013. The Contribution of Benefit-in-Kind Taxation Policy in Britain to the 'Peak Car' Phenomenon. *Transport Reviews* 33 (5): 526–547.

Lee-Gosselin, M.E.H. 2017. Beyond 'Peak Car': A Reflection on the Evolution of Public Sentiment About the Role of Cars in Cities. *IATSS Research* 40 (2): 85–87. http://linkinghub.elsevier.com/retrieve/pii/S0386111216300115.

Lin, J., and L. Long. 2008. What Neighborhood Are You In? Empirical Findings of Relationships Between Household Travel and Neighborhood Characteristics. *Transportation* 35 (6): 739–758. http://link.springer.com/10.1007/s11116-008-9167-7.

Lucas, K., B. van Wee, and K. Maat. 2015. A Method to Evaluate Equitable Accessibility: Combining Ethical Theories and Accessibility-Based Approaches. *Transportation* 43: 473.

Manville, M., D.A. King, and M.J. Smart. 2017. The Driving Downturn: A Preliminary Assessment. *Journal of the American Planning Association* 83 (1): 42–55. https://www.tandfonline.com/doi/full/10.1080/01944363.2016.1247653.

Martin, A., J. Panter, M. Suhrcke, and D. Ogilvie. 2015. Impact of Changes in Mode of Travel to Work on Changes in Body Mass Index: Evidence from the British Household Panel Survey. *Journal of Epidemiology and Community Health* 69 (8): 753–761. http://jech.bmj.com/lookup/doi/10.1136/jech-2014-205211.

McDonald, N.C. 2015. Are Millennials Really the 'Go-Nowhere' Generation? *Journal of the American Planning Association* 81 (2): 90–103. http://www.tandfonline.com/doi/full/10.1080/01944363.2015.1057196.

McIntosh, J., R. Trubka, J. Kenworthy, and P. Newman. 2014. The Role of Urban Form and Transit in City Car Dependence: Analysis of 26 Global Cities from 1960 to 2000. *Transportation Research Part D: Transport and Environment* 33: 95–110. http://linkinghub.elsevier.com/retrieve/pii/S136192091400114X.

Minton, J., L. Vanderbloemen, and D. Dorling. 2013. Visualizing Europe's Demographic Scars with Coplots and Contour Plots. *International Journal of Epidemiology* 42 (4): 1164–1176. http://ije.oxfordjournals.org/content/42/4/1164.full.

SACTRA (Standing Advisory Committee on Trunk Road Assessment). 1994. *Trunk Roads and the Generation of Traffic*. London: HMSO.

Schöley, J., and F. Willekens. 2017. Visualizing Compositional Data on the Lexis Surface. *Demographic Research* 36: 627–658. http://www.demographic-research.org/volumes/vol36/21/.

Serafino, P., and R. Tonkin. 2014. Intergenerational Transmission of Disadvantage in the UK and EU. Office for National Statistics. http://webarchive.nationalarchives.gov.uk/20160107123147; http://www.ons.gov.uk/ons/dcp171766_378097.pdf.

Sharmeen, F., T. Arentze, and H. Timmermans. 2014. An Analysis of the Dynamics of Activity and Travel Needs in Response to Social Network Evolution and Life-Cycle Events: A Structural Equation Model. *Transportation Research Part A: Policy and Practice* 59: 159–171. http://linkinghub.elsevier.com/retrieve/pii/S0965856413002322.

St-Louis, E., K. Manaugh, D. van Lierop, and A. El-Geneidy. 2014. The Happy Commuter: A Comparison of Commuter Satisfaction Across Modes. *Transportation Research Part F: Traffic Psychology and Behaviour* 26: 160–170. http://linkinghub.elsevier.com/retrieve/pii/S1369847814001107.

Stokes, G. 2013. The Prospects for Future Levels of Car Access and Use. *Transport Reviews* 33 (3): 360–375. http://www.tandfonline.com/doi/abs/10.1080/01441647.2013.800614.

Teubner, T., and C.M. Flath. 2015. The Economics of Multi-Hop Ride Sharing. *Business & Information Systems Engineering* 57 (5): 311–324. http://link.springer.com/10.1007/s12599-015-0396-y.

Thomas, G.O., and I. Walker. 2015. Users of Different Travel Modes Differ in Journey Satisfaction and Habit Strength but Not Environmental Worldviews: A Large-Scale Survey of Drivers, Walkers, Bicyclists and Bus Users Commuting to a UK University. *Transportation Research Part F:*

Traffic Psychology and Behaviour 34: 86–93. http://linkinghub.elsevier.com/retrieve/pii/S1369847815001205.

Thomopoulos, N., and M. Givoni. 2015. The Autonomous Car—A Blessing or a Curse for the Future of Low Carbon Mobility? An Exploration of Likely vs. Desirable Outcomes. *European Journal of Futures Research* 3 (1): 14. http://link.springer.com/10.1007/s40309-015-0071-z.

Tilley, S., and D. Houston. 2016. The Gender Turnaround: Young Women Now Travelling More Than Young Men. *Journal of Transport Geography* 54: 349–358. http://linkinghub.elsevier.com/retrieve/pii/S0966692316303581.

Tyrovolas, S., E. Polychronopoulos, M. Morena, A. Mariolis, S. Piscopo, G. Valacchi, V. Bountziouka, F. Anastasiou, A. Zeimbekis, D. Tyrovola, A. Foscolou, E. Gotsis, G. Metallinos, G. Soulis, J.-A. Tur, A. Matalas, C. Lionis, L.S. Sidossis, and D. Panagiotakos. 2017. Is Car Use Related with Successful Aging of Older Adults? Results from the Multinational Mediterranean Islands Study. *Annals of Epidemiology* 27 (3): 225–229. http://linkinghub.elsevier.com/retrieve/pii/S1047279717300285.

Uhrig, S.C.N. 2008. The Nature and Causes of Attrition in the British Household Panel Survey. *Essex*. https://www.iser.essex.ac.uk/files/iser_working_papers/2008-05.pdf.

Urry, J. 2004. The 'System' of Automobility. *Theory, Culture & Society* 21 (4–5): 25–39. http://journals.sagepub.com/doi/10.1177/0263276404046059.

Vaupel, J.W., Z. Wang, K. Andreev, and A.I. Yashin. 1997. *Population Data at a Glance: Shaded Contour Maps of Demographic Surfaces Over Age and Time*. Denmark: University Press of Southern Denmark. http://www.abebooks.co.uk/servlet/BookDetailsPL?bi=2944819605.

Yap, M.D., G. Correia, and B. van Arem. 2016. Preferences of Travellers for Using Automated Vehicles as Last Mile Public Transport of Multimodal Train Trips. *Transportation Research Part A: Policy and Practice* 94: 1–16. http://linkinghub.elsevier.com/retrieve/pii/S0965856416307765.

3

Mobility and Ageing: A Review of Interactions Between Transport and Technology from the Perspective of Older People

Kate Pangbourne

Introduction

This chapter is an overview of the issues for older people in relation to the nexus between transport and technology, that is intended to set a context to complement the other contributions to this volume particularly Chapters 5 and 9 Dowds et al. and Fitt. A key aim of this chapter is to demonstrate that mobility and technology are intertwined in complex ways, and that even non-transport technologies may impact older people's experience and achievement of mobility. Understanding the nexus between mobility, information and communication technologies (ICT) and older people can help us design accessible and acceptable technologies to support well-being and health in older age. This matters because new ICT is increasingly being relied upon to support service delivery in both the public and private sectors. For example, ICT is increasingly

K. Pangbourne (✉)
Institute for Transport Studies, University of Leeds, Leeds, UK
e-mail: K.J.Pangbourne@leeds.ac.uk

© The Author(s) 2018
A. Curl and C. Musselwhite (eds.), *Geographies of Transport and Ageing*,
https://doi.org/10.1007/978-3-319-76360-6_3

harnessed to support centralisation of aspects of service provision, or even to virtualise it entirely. In many cases, this removes, or modifies, the need to travel to access certain services. For example, personal banking can be carried out online, from transferring money between accounts to paying bills. This removes the need to travel to a branch, and as a result of increasing uptake, banks have rationalised and reduced their branch networks. As a society, this should make us ask the question 'How does this impact on people, and what are their perceptions of this shift?' If we find that particular groups within society, such as older people, are significantly negatively impacted, we should consider if there are good societal reasons to mitigate these effects in some way. Indeed, in the UK some banks are making some efforts to engage different groups with new technology, for example Barclays Bank has a 'Digital Eagles' scheme, for which older adults are a key audience.[1] In relation to the development of new technologies in general, it is recognised that older people may be more likely to experience a reluctance or difficulty in adopting the technologies. However, being over 65 is not in itself a predictor of low technology adoption. The older age groups are heterogeneous, with different levels of income and education affecting adoption rates. The greater prevalence of cognitive and physical impairments can also impact on technology adoption. Attitudes are also relevant, as some older people may not see a need to adopt technologies that they have managed many decades without (Smith 2014).

Conversely, there are opportunities to utilise ICT to improve the experience of service access specifically for older people. Healthcare is one domain where ICT is increasingly used to support older people, for example through assistive technologies, the majority of which are not intended to support out-of-home mobility, but to support ageing-in-place and reduce the need for travel to health centres. Where travel to health centres and hospitals is still necessary, centralisation of service provision in the health sector, and loss of public transport in the transport sector have had the twin effects of creating a great burden for people of any age experiencing ill-health, but it is particularly

[1] http://www.barclays.co.uk/digital-confidence/eagles/.

problematic for older adults who cannot access a car. ICT could be used to dovetail appointment times with available transport, but this type of change needs to be led and delivered by the healthcare provider, ideally in collaboration with transport providers.

ICT and other technologies are converging rapidly in the transport sector, enabling almost real-time access to demand responsive services, which have existed for some decades with pre-booking first by telephone, but with online options introduced later. This transport-ICT convergence is resulting in the emergence of many new services based on the affordances of smartphone technology. However, older people are not usually regarded as the target market for these products—Alba (2016) deduces from smartphone sales trends that the market is more or less saturated such that everyone who wants one has one, and by further inference from the demographics of smartphone ownership, older people own them at lower rates than younger people (e.g. Pew Research Center 2017). There are a number of reasons why smartphone adoption is lower amongst older people—new technology adoption in general slows with age (as mentioned above), and it is clear to see that most advertising of new technology products portray images of youth, unless it is a technology designed mainly for older people (such as stair lifts or mobility scooters). Another key reason is likely to be design or usability, particularly in relation to the touch screen-based user interface, which poses particular issues for people with visual or upper limb impairments (Mi et al. 2014). Conversely, older people are often described as a key beneficiary of a game-changing technology which is on the brink of an innovation breakthrough. For example, autonomous, or self-driving vehicles (SDV), are described by some authors as a key assistive technology for maintaining out-of-home mobility.

The remainder of this chapter is structured by a brief reprise of the importance of out-of-home mobility for healthy ageing in the context of an ageing global population, touching on the heterogeneity of needs and wants amongst older people. The next section will examine aspects of the nexus between mobility, ICT and older people in order to show how well-being and health can be supported using technologies that are accessible and accepted by older people, with subsections on mobility technologies and possible future developments. The final section before the conclusions and policy recommendations considers the potential role of age-friendly design in transport policy and cities. The perspective is primarily one from the Global North.

The Importance of Out-of-Home Mobility for Healthy Ageing

> The ability to move about—and by extension to travel—is required to navigate from point A to point B, to seek out places of subjective interest or that are essential to meeting daily material needs, to participate in cultural and recreational activities, and to maintain social relations, familiar habits, and life styles—in short, to live an autonomous life for as long as one's mental and physical capacities permit one to participate actively in society (Schaie, 2003). At the same time, age-related changes such as physical, cognitive, and/or sensory impairments and social losses may limit older adults' possibilities of ambulating and venturing out. (Mollenkopf et al. 2017, p. 267)

As is evident from the quotation from Mollenkopf et al. (2017) above, mobility is a crucial issue for healthy ageing, and we are we are reaching greater ages at unprecedented rates. It is forecast that the proportion of the global population of 80 years or older will be 20% by 2050 (UN 2015). Setting that into the context of a global population that is growing at 1% per annum, the growth rate in the 80+ age group is 4% (HelpAge International 2012 cited by Sixsmith 2013). Geographically, more than 60% of the older population are in less developed regions, with the 2050 forecast being 80%. This poses a considerable challenge for service providers in supporting older people to extend good health and quality of life for as long as possible. The concept of 'active ageing' is increasingly dominating the policy discourse, defined by the World Health Organization as:

> Active ageing is the process of optimizing opportunities for health, participation and security in order to enhance quality of life as people age … It allows people to realize their potential for physical, social, and mental

> well being throughout the life course and to participate in society according to their needs, desires and capacities, while providing them with adequate protection, security and care when they require assistance. (WHO 2002, p. 12)

Active ageing spans many policy areas, including health, education, and housing, though mobility underpins all of them (Johnson et al. 2017), as the key individual and policy goal is perceived as being maintaining autonomy and independence for as long as possible. Somewhat confusingly however, the Active Ageing Framework also talks of interdependence and intergenerational giving (WHO 2002).

Bodily mobility is more likely to be compromised in older adults, making the achievement of out-of-home mobility more challenging. However, there is a sizeable body of research that demonstrates that out-of-home mobility is an important determinant of quality of life for older adults (Metz 2000; Spinney et al. 2009; Gilhooly et al. 2002), where quality of life includes aspects such as being autonomous and having a social life (Ziegler and Schwanen 2011), being able to obtain daily necessities and healthcare, and be part of a community (Kaiser 2009), which collectively are associated physical and mental benefits that we might term 'well-being' (Reardon and Abdallah 2013; Simonsick et al. 2005). In summary therefore, mobility, and especially out-of-home/outdoor mobility, has clearly been identified to be a key factor in successful ageing for older adults (Mollenkopf 2005; Kaspar et al. 2015), due to the powerful effect of a sense of fulfilment that is conferred by independence (Mokhtarian et al. 2015; Musselwhite 2011).

What factors influence the levels of out-of-home mobility that older adults are able to achieve? Numerous research studies have quantified the out-of-home mobility of older people such as those cited by Mollenkopf et al. (2017) who note that whilst there are differences explainable through national 'peculiarities', there are general tendencies.

Firstly, the amount of travel undertaken by older adults has been increasing over the last twenty years. However, within the 'older adult' category, the amount of travel is significantly reduced with increasing age. This is not surprising, as it is mainly due to a decline in health and an increase in sensory impairment. Secondly, in relation to a key transport concern of mode choice, those with a driving licence and access to a car will tend to travel more than those who do not have such access. There is thus a gender effect, as the current cohort of older women has a lower level of education, a lower income and is less likely to have a driving licence than men of the same age. Thirdly, therefore, older women use public transport more than older men, who use the car more often, travelling more often and further (see, e.g., Banister and Bowling 2004; Marottoli et al. 1997; Mollenkopf et al. 2004; Rosenbloom 2004).

As possession of driving licence and access to a car is a significant predictor of higher overall mobility, the effect of ill-health and sensory impairment resulting in driving cessation can be significant (Zeitler and Buys 2015; Souders and Charness 2014). Maratolli et al. (2000) found a strong association between driving cessation and decreased out-of-home activity, even after correcting for socio-demographic and health-related factors. In a highly car-dependent society, a lack of alternatives to driving leads to transport disadvantage amongst older adults (Engels and Liu 2011). This is juxtaposed against a normative expectation of increased mobility amongst older adults (Alsnih and Hensher 2003). Consequently, a great deal of effort is expended on extending safe driving for longer, through various, technological means, in order to prolong independent living (Nordbakke and Schwanen 2015).

In contrast to these efforts to maintain driving as a component of independent living, there are parallel developments, largely from the health sector, which focus on ICT as a technological support for independent living that bypasses a need for 'out-of-home' mobility.

Mobility, ICT and Healthy Old Age

ICT has emerged as a major strand in research and development for older people, as it is perceived as being able to provide better 'care' at potentially lower cost. Health-related ICT, or e-health, is largely intended to deliver remote health monitoring of older people living in the community ('care at a distance', O'Hanlon et al. 2012) rather than in residential care, but it overlaps quite heavily with 'efficiency' measures in healthcare provision, through reducing the need for human-delivered care. Remote health monitoring is now quite highly advanced, and can involve contactless sensors capable of detecting falls, or providing continuous monitoring of important indicators such as blood pressure (Malasinghe et al. 2017). Whilst the motive is one of benefit for the patient in being able to remain in personal environments, there is also a cost advantage to avoiding hospital stays:

> With the new remote health monitoring applications, elderly patients can engage in daily activities without support from a caretaker. So, these applications support activities like sitting, standing, using the bathroom, watching television, reading and sleeping, with least inconvenience to the user. Even if there are wearable sensors, these pose minimum effect to the activities. One such example is smart wrist-watch based sensors. (Malasinghe et al. 2017, p. 1)

However, this is not necessarily a perfect solution, for as with the inexorable shift online of banking services, e-health also potentially reduces the amount of social contact and out-of-home mobility that older people experience.

However, Sixsmith (2013) points out that the majority of assistive technology and e-health research is focused on assisting those who are already impaired in some way, when it is actually the case that the majority of older adults are currently healthy and active. He recommends that technology Research and Development should focus on providing this group with products and services that maintain their proactive and independent status for as long as possible. Thus, wearable technologies that encourage physical activity such as Fitbit, IQ-FIT, Moov Now or Garmin could perhaps be promoted more specifically to the older adult market. Research suggests that they could be useful for older adults by providing them with data for self-monitoring and to encourage greater levels of activity to support health (O'Brien et al. 2015; Lyons et al. 2017).

The social and political tendency to stereotype older people as passive and dependent overlooks the fact that most older people are proactive agents living healthy and independent lives, and the goal should be to support this status, rather than expecting that they will become passive receivers of care (notwithstanding the increased proportion of self-reported ill-health after 75 years). It is increasingly recognised that old age is not a homogenous category (Haustein 2012). In terms of health, this variety means that 'chronological age is not a relevant marker for understanding, measuring, or experiencing healthy aging' (Lowsky et al. 2014, p. 640).

Technology Take-up Amongst Older People

Attitudes to technology are similarly heterogeneous. Interface design studies from the turn of the twenty-first century demonstrated that older technology users are much more diverse than younger and middle-aged users (Gregor et al. 2002). In a survey of ICT use amongst the over-60s in England and Wales, Selwyn et al. (2003) found that part of the reason for the low usage of ICT in this age group, despite the growing numbers of 'silver surfers', is to do with 'relevance'. In essence, ICT was used if it had relevant function or content for the user. Other literature also suggested that many older people are not technophobe per se but see little relevance in the use of digital technologies in their daily lives, despite the 'information society' rhetoric and advances in individualisation (Pangbourne et al. 2010).

The MOBILATE study conducted a survey across 5 European countries of senior people's mobility, and included some data on technology use and acceptance. At that time, 2005, the share of use of the technologies studied was low amongst the over 55s, but these older people's experiences of common technologies like cash machines were quite positive, though public transport ticket dispensers were less well received, and commonly regarded as excessively complicated. It was noted that a high educational level, high income and good health were all factors that contributed to older adults being able to overcome barriers to use (Tacken et al. 2005).

As described in the introduction, older people are under considerable pressure to adopt new technologies across their lives: services as diverse as government, banking, insurance, transport ticketing, healthcare services and shopping are increasingly shifting to web-based and mobile interfaces, with physical branches and telephone call centres increasingly less available or frustrating to access. However, at the time of the MOBILATE study most people over the age of 55 were not habitual users of PCs in the workplace, and have thus had much less opportunity to become familiar with the conventions of human-computer interfaces. The spread of smartphones has introduced new conventions and affordances for those who adopt them.

New generations continuously enter the senior age brackets, and increasingly have familiarity with at least some ICTs. Nevertheless, adoption and acceptance cannot be taken for granted—many older people are suspicious of contactless and mobile payment methods for example, and are also more guarded (yet less skilled) when it comes to issues of personal data privacy and location tracking services. It is not clear whether this is a cohort effect or a reflection of some of the cognitive effects of ageing. However, if the technology adoption rate of mobile services matches that of the take-up of smartphones, internet and broadband amongst older people, then it is most likely a cohort effect that is fading—see, for example, recent statistics from the United States, where older adults up to 69 have similar rates of internet and broadband use as the general population, though smartphone adoption is more strongly linked to income (Anderson and Perrin 2017).

The newer interfaces, often with new business models or concepts such as multi-layer menu systems, are harder to adopt for older people than for younger people, whilst due at least in part to a general 'slowing down' of cognitive ability, lack of uptake is more likely to relate to an attitudinal reluctance to keep learning new things, or a view that there is really no need for novelty, unless the ICT supports hobbies or other interests that increase individual motivation to overcome the learning hurdles that everyone experiences. There are also practical difficulties for older people—eyesight increasingly becomes a factor in using smartphones and items with screens and keyboards. MOBILATE found that increasing age and being female were associated with lower levels of technology adoption, notwithstanding some interesting reversals, as women used ticket dispensing machines more than men. This is assumed to be because, as more frequent users of public transport, they have more knowledge of how to use public transport, including how to buy tickets. Nevertheless, MOBILATE's data show that gender and age are important predictors for the use of new technology (Tacken et al. 2005, p. 131).

Older People and Technologies for Mobility

It is important to remember that with the exception of unaided walking,[2] all our out-of-home mobility is facilitated by technologies of varying degrees of sophistication. To date, the technological innovation in transport with the largest impact has to be the emergence and rapid dominance of the internal combustion engine, particularly in how it has provided personal mobility to millions through the car. The unique features of automobiles have totally transformed the organisation of society wherever car ownership has taken hold. Because of 'automobility's exceptional power to remake time-space, especially because of its peculiar combination of flexibility and coercion' (Urry 2004, p. 27), our societal preconceptions of mobility in older age are inevitably underpinned by the general expectation that the car is the main provider of personal mobility for most of our lives. The car also underpins the generally increased expectation of mobility as highlighted above. In other words the invention of the car totally reconfigured social practices, urban layout and the distribution of land uses, which helps to explain the great focus in the mobility and ageing literature on the impact of driving cessation as a significant event for older people, carrying with it a number of negative connotations and impacts in contexts where vital services are dislocated in time and space from homes (as touched on above).

Nevertheless, it is intuitive to expect that other transport technologies that support out-of-home mobility[3] will become a bigger issue in

[2]Some would contend that shoes are also technology and certainly orthotics can be added to footwear as assistive technologies.

[3]It should be noted that most of the gerontology and health literature conceptualise 'mobility' as an embodied capacity, i.e. the ability to walk, and perform independent actions, within the home, rather than as an act of travel, and mobility assistive technologies are wheelchairs, walkers, scooters, etc.

relation to an ageing population. At the same time, ICT innovation is having as rapid and profound an impact on the shape of transport provision as it is in healthcare. Yet as discussed above, these innovations tend to achieve a slower rate of uptake amongst older people than younger. What does this imply for utilising ICT in providing future mobility services for older people?

In a focus group study, Pangbourne et al. (2010) (Fig. 3.1) demonstrated that the concerns of older people in relation to interactions between transport, healthcare and ICT are personal (the need for services to address feelings of vulnerability, a perception of a loss of social contact where ICT is used, but recognition that ICT can support individualisation), that there are informational requirements (older people want to be consulted about changes to services, they want healthcare information online, and they want communication about bookings and transport to healthcare appointments). Associated with these perceptions is an awareness that accessibility (to real-time travel information, transport and appointment times not matching up, having online appointment booking facilities) is not assured, and that usability of interfaces need attention. These issues emerged through discussion that initially appeared to be about reliability of transport services, parking at

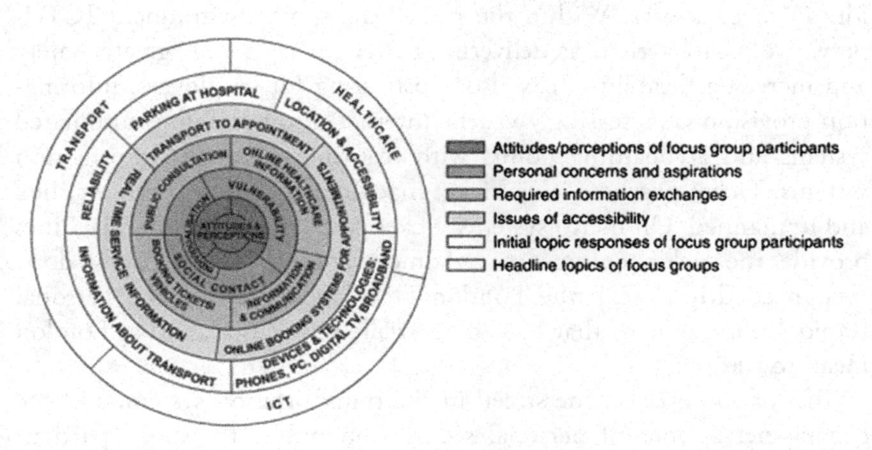

Fig. 3.1 Focus group findings regarding intersections between transport, health care and ICT raised by older people (Pangbourne et al. 2010, p. 322)

hospitals, locations of healthcare facilities, and ICT (hardware and software) that they use.

Against this background, discussion of the role of ICT technologies for supporting mobility amongst older people tends to be that which somehow compensates them for their 'diminished' life experience in relation to getting out and about. ICT aimed at older people is seen often seen as mitigation and adaptation rather than a positive development. However, there are now developments to support those with cognitive impairments to continue enjoying outdoor mobility by using GPS trackers (Kaspar et al. 2015), as it has been shown that outdoor mobility is a critical factor in healthy ageing. Careful use of ICT is only one factor: social practices, and design of the built environment (e.g. buildings with heavy external doors can 'trap' mobility-impaired people inside, rendering them dependent on another person to be able to get outside), also need age-friendly evolution (Rantanen 2013).

ICT, Transport and Future Developments in Mobility

Within the transport environment itself, ICT is developing quite rapidly in diverse ways. Within the public transport environment ICT is pervasive, characterised as delivering better management, greater safety and increased flexibility. It is also substituting for employees: information provision is increasingly via the internet through online automated systems and at boarding points with real-time passenger information systems. Ticket purchase is via ticket machines at stations both manned and unmanned. On metro systems, ticket gates and validation machines provide the ticket enforcement. Some transport services now don't even need drivers (e.g. the London Docklands Railway, the Personal Rapid Transit systems that have been trialled in Masdar and at London Heathrow airport).

The loss of 'eyes on the street' in the transportation system, and the general perceptions of personal security on public transport, particularly for older people, is an issue which has attracted some attention in research. For example, Sochor (2013) has investigated whether Swedish older adults find a range of public transport ICTs reassuring (video surveillance in public transportation, real-time travel information, and a personal, pedestrian navigation system with public transportation information). Perceptions were neutral regarding privacy and positive in support a sense of assurance, especially for women, who are known to feel more vulnerable in certain mobility settings, though men were more interested in the 'technology'. Surprisingly, personal control over the ICT (in the case of a navigation system with information) did not enhance the sense of assurance. In a further development of this work, Sochor and Nikitas (2016) added evidence from Britain to a Swedish study of the technology perceptions of visually impaired people. They conclude that whilst the attitudes of older people are generally accepting, they are prosocial in considering the benefits rather than personal, as sometimes the technologies are perceived as 'complicated'. They conclude that these technologies are only one element in a complex sociotechnical system that is challenging for meeting the needs of older people.

There are a number of vehicle developments that are sometimes portrayed as likely to be beneficial for older adults. The key innovation, which is not yet market ready, is the emergence of self-driving vehicles (SDV), also known as autonomous vehicles. Therefore, it is difficult to evaluate the likely outcomes for older people. On the face of it, SDV could be regarded as an assistive technology, and inevitably one that would have quite a radical restructuring impact on social practices and other transport modes (Shergold et al. 2015). Shergold (2016) also review what is currently known about the impacts of SDVs and driver assistance technologies, and conclude that the benefits of this group of technologies is likely to have particular benefits for older people at risk of losing independence as a result of driving cessation.

However, in order to ensure that any technology, including the built environment itself, is able to support the mobility and independence of older people, attention needs to be paid to design issues (e.g. I referred above to the particular difficulties posed by smartphone touch screens). In the next section, consideration is given to age-friendly design and policy.

Age-Friendly Design and Transport Policy

Rantanen (2013) described evocatively how the design choice of a heavy front door can impede an older person's outdoor mobility (see above). At the larger scale, transport and the built environment is dictated by a complex web of urban and transport policies. Johnson et al. (2017) have reviewed national approaches to older adult's transport needs in Europe. Drawing on their own literature review, and on the TRACY Project (2012) they identify 11 qualities of an age-friendly transport system (Affordability, Availability, Barrier-free, Comfortable, Comprehensible, Efficient, Friendly, Reliable, Safe, Secure and Transparent). Aguiar and Macário (2017) also highlight the need for mobility policy to be more focused on the needs of an older population, highlighting important infrastructure measures that may seem trivial, such as pavement improvements, that have very positive benefits for older people. In a similar vein, the GLIDE project in Singapore trialled technologies to provide longer crossing times for older or physically challenged pedestrians (Debnath et al. 2011). These age-friendly, active ageing initiatives and criteria are entirely compatible with the World Health Organization's 'age-friendly cities' objective, which highlights that by means of inclusive design, all age groups benefit (WHO 2007). However, in Johnson et al's (ibid) study, their findings suggest that the most consistent qualities that are addressed by government are safety, barrier freedom and affordability. They conclude that the emphasis placed on these qualities is due to their tangibility and relative ease of implementation rather than any objective assessment of likely benefit. The neglect of the softer intangibles has attracted some research attention (e.g. Hounsell et al. 2016; Grotenhuis et al. 2007 or O'Neill 2016), but has yet to gain much traction in implementation.

Conclusions

Whilst the technologies that are being advanced in both e-health and transport are exciting and could be of significant benefit to society, the predominant government and industry discourse carries a clear 'technology optimism bias' which colours thinking about the capabilities/benefits of all technological innovations, including ICTs. Of course for the industry, this comes from the fact that they are marketing their products. Governments, presumably, are marketing themselves as re-electable and wish to be perceived as forward thinking or facilitative of commerce. A lot of the ICT initiatives at the nexus of Smart Mobility and Smart Cities can only be implemented in cities that have high quality infrastructure already in place, and there are many areas (urban, peri-urban and rural in the both the Global North and Global South) where the telecommunications infrastructure is simply behind the curve, and where the most useful ICTs cannot yet be deployed, even quite basic real-time passenger information. This is a serious shortcoming that affects everyone, not only older people. In rural areas, where populations are predominantly older, there is clearly a significant issue that needs to be addressed, as many transport innovations will be unavailable or unaffordable in 'unconnected' areas.

As we have seen, Johnson et al.'s (2017) analysis suggests that both research and policy have thus far focused on objective and measurable qualities (safety, barrier freedom and affordability in particular), neglecting the softer, more subjective qualities of the lived experience of older people. Increasingly, research is showing that the subjective and context-related factors are very significant in suppressing the mobility of older people, whilst at the same time, getting out and about independently is increasingly understood as a crucial issue in supporting health and well-being.

Thus qualities such as security (in the sense that transport provision is perceived as secure, and addresses confidence issues that older people may have), friendliness and comfort should be more directly addressed. Significantly, Johnson et al. (2017) also talk of the single-mode specificity of most of the policy documents they evaluated. This neglects the need to join up transport policies to improve age-friendliness. For example, having policies which limit driving licences for older people are essential for safety reasons, but without providing alternatives with the right age-friendly qualities, those affected are doomed to profound and damaging immobility.

Policy and Research Recommendations

Society needs to ask critical questions—are mobility technology changes such as a wholesale transition to self-driving vehicles worth striving for or is remote service provision, including healthcare a better option? More probably there should be a blend of approaches, given the role that out-of-home mobility plays in maintaining healthy physical and mental outlooks at any age, though particularly in older age. The introduction of new technologies, whether transport or health-related, will shape future societies and the lived experience of older people through subtle impacts on social practices. However, it is good to remember the words of the World Health Organization, in its age-friendly cities guidance document:

> Because active ageing is a lifelong process, an age-friendly city is not just "elderly friendly". Barrier-free buildings and streets enhance the mobility and independence of people with disabilities, young as well as old. Secure neighbourhoods allow children, younger women and older people to venture outside in confidence to participate in physically active leisure and in social activities. Families experience less stress when their older members have the community support and health services they need. The whole community benefits from the participation of older people in volunteer or paid work. Finally, the local economy profits from the patronage of older adult consumers. The operative word in age-friendly social and physical urban settings is enablement. (WHO 2007, p. 6)

References

Aguiar, B., and R. Macário. (2017). The Need for an Elderly Centred Mobility Policy. *Transportation Research Procedia* 25: 4355–4369.

Alba, D. 2016. It's Official: The Smartphone Market Has Gone Flat. *Wired*. https://www.wired.com/2016/04/official-smartphone-market-gone-flat/. Accessed 31 October 2017.

Alsnih, R., and D.A. Hensher. 2003. The Mobility and Accessibility Expectations of Seniors in an Aging Population. *Transportation Research Part A: Policy and Practice* 37 (10): 903–916.

Anderson, M., and A. Perrin. 2017. *Technology Use Among Seniors*. Pew Research Center. http://www.pewinternet.org/2017/05/17/technology-use-among-seniors/. Accessed 6 September 2017.

Banister, D., and A. Bowling. 2004. Quality of Life for the Elderly: The Transport Dimension. *Transport Policy* 11: 105–115.

Boschmann, E.E., and S.A. Brady. 2013. Travel Behaviors, Sustainable Mobility, and Transit-Oriented Developments: A Travel Counts Analysis of Older Adults in the Denver, Colorado Metropolitan Area. *Journal of Transport Geography* 33: 1–11.

Broome, K., K. McKenna, J. Fleming, and L. Worrall. 2009. Bus Use and Older People: A Literature Review Applying the Person–Environment–Occupation Model in Macro Practice. *Scandinavian Journal of Occupational Therapy* 16 (1): 3–12.

Broome, K., E. Nalder, L. Worrall, and D. Boldy. 2010. Age-Friendly Buses? A Comparison of Reported Barriers and Facilitators to Bus Use for Younger and Older Adults. *Australasian Journal on Ageing* 29 (1): 33–38.

Debnath, A.K., M. Haque, and H. Chin. 2011. Sustainable Urban Transport: Smar Technology Initiatives in Singapore. *Transportation Research Record* 2243: 38–45.

Engels, B., and G.J. Liu. 2011. Social Exclusion, Location and Transport Disadvantage Amongst Non-Driving Seniors in a Melbourne Municipality, Australia. *Journal of Transport Geography* 19 (4): 984–996.

Gilhooly, M., K. Hamilton, M. O'Neill, J. Gow, N. Webster, F. Pike, and D. Bainbridge. 2002. *Transport and Ageing: Extending Quality of Life for Older People via Public and Private Transport*.

Gregor, P., A.F. Newell, and M. Zajicek. 2002. Designing for Dynamic Diversity: Interfaces for Older People. *ASSETS '02 Proceedings of the Fifth International ACM Conference on Assistive Technologies*, 151–156.

Grotenhuis, J.-W., B.W. Wigmans, and P. Rietveld. 2007. The Desired Quality of Integrated Multimodal Travel Information in Public Transport: Customer Needs for Time and Effort Savings. *Transport Policy* 14: 27–38.

Haustein, S. 2012. Mobility Behavior of the Elderly: An Attitude-Based Segmentation Approach for a Heterogeneous Target Group. *Transportation* 39 (6): 1079–1103.

Hounsell, N.B., B.P. Shrestha, M. McDonald, and A. Wong. 2016. Open Data and the Needs of Older People for Public Transport Information. *Transport Research Procedia* 14: 4334–4343.

Johnson, R., J. Shaw, J. Berding, M. Gather, and M. Rebstock. 2017. European National Government Approaches to Older People's Transport System Needs. *Transport Policy* 59: 17–27.

Kaiser, H.J. 2009. Mobility in Old Age: Beyond the Transportation Perspective. *Journal of Applied Gerontology* 28: 411–418.

Kaspar, R., F. Oswald, H.W. Wahl, E. Voss, and M. Wettstein. 2015. Daily Mood and Out-of-Home Mobility in Older Adults: Does Cognitive Impairment Matter? *Journal of Applied Gerontology* 34: 26–47.

Lowsky, D.J., S.J. Olshansky, J. Bhattacharya, and D.P. Goldman. 2014. Heterogeneity in Healthy Aging. *The Journals of Gerontology Series A: Biological Sciences and Medical Sciences* 69 (6): 640–649.

Lyons, E.J., M.C. Swartz, Z.H. Lewis, E. Martinez, and K. Jennings. 2017. Feasibility and Acceptability of a Wearable Technology Physical Activity Intervention with Telephone Counseling for Mid-Aged and Older Adults: A Randomized Controlled Pilot Trial. *JMIR mHealth and uHealth* 5 (3): Online only.

Malasinghe, L.P., N. Ramzan, and K.J. Dahal. 2017. Remote Patient Monitoring: A Comprehensive Study. *Journal of Ambient Intelligence and Humanized Computing*, 1–20. https://doi.org/10.1007/s12652-017-0598-x.

Marottoli, R.A., C.F.M. de Leon, T.A. Glass, C.S. Williams, L.M. Cooney Jr., L.F. Berkman, and M.E. Tinetti. 1997. Driving Cessation and Increased Depressive Symptoms: Prospective Evidence from the New Haven EPESE. Established Populations for Epidemiological Studies of the Elderly. *Journal of the American Geriatrics Society* 45: 202–206.

Marottoli, R.A., C.F.M. de Leon, T.A. Glass, C.S. Williams, L.M. Cooney Jr., and L.F. Berkman. 2000. Consequences of Driving Cessation Decreased Out-of-Home Activity Levels. *The Journals of Gerontology Series B: Psychological Sciences and Social Sciences* 55 (6): S334–S340.

Metz, D.H. 2000. Mobility of Older People and Their Quality of Life. *Transport Policy* 7 (2): 149–152.

Mi, N., L.A. Cavuoto, K. Benson, T. Smith-Jackson, and M.A. Nussbaum. 2014. A Heuristic Checklist for an Accessible Smartphone Interface Design. *Universal Access in the Information Society* 13: 351–365.

Mokhtarian, P., I. Salamon, and M. Singer. 2015. What Moves Us? An Interdisciplinary Exploration of the Reasons for Traveling. *Transport Reviews* 35: 250–274.

Mollenkopf, H. 2005. *Enhancing Mobility in Later Life: Personal Coping, Environmental Resources and Technical Support; The Out-of-Home Mobility of*

Older Adults in Urban and Rural Regions of Five European Countries, vol. 17. Amsterdam, The Netherlands: IOS Press.

Mollenkopf, H., A. Heiber, and H.W. Wahl. 2017. Continuity and Change in Older Adults' Out-of-Home Mobility Over Ten Years: A Qualitative-Quantitative Approach. In *Knowledge and Action, Knowledge and Space*, ed. P. Meusburger et al., Chapter 15, 267–289.

Mollenkopf, H., F. Marcellini, I. Ruoppila, Z. Széman, M. Tacken, and H.W. Wahl. 2004. Social and Behavioural Science Perspectives on Out-of-Home Mobility in Later Life: Findings from the European Project MOBILATE. *European Journal of Ageing* 1 (1): 45–53.

Musselwhite, C.B.A. 2011. Successfully Giving Up Driving for Older People.

Nordbakke, S., and T. Schwanen. 2015. Transport, Unmet Activity Needs and Wellbeing in Later Life: Exploring the Links. *Transportation* 42: 1129–1151.

O'Brien, T., M. Troutman-Jordan, D. Hathaway, S. Armstrong, and M. Moore. 2015. Acceptability of Wristband Activity Trackers Among Community Dwelling Older Adults. *Geriatr Nurs.* 36 (2 Suppl.): S21–S25.

O'Hanlon, S., A. Bourke, and V. Power. 2012. E-Health for Older Adults. Engaging Older Adults with Modern Technology: Internet Use and Information Access Needs: Internet Use and Information Access Needs, 229.

O'Neill, D. 2016. Towards an Understanding of the Full Spectrum of Travel-Related Injuries Among Older People. *Journal of Transport and Health* 3: 21–25.

Pangbourne, K., P.T. Aditjandra, and J.D. Nelson. 2010. New Technology and Quality of Life for Older People: Exploring Health and Transport Dimensions in the UK context. *IET Intelligent Transport Systems* 4 (4): 318–327.

Pew Research Center. 2017. *Mobile Fact Sheet*. http://www.pewinternet.org/fact-sheet/mobile/. Accessed 31 October 2017.

Rantanen, T. 2013. Promoting Mobility in Older People. *Journal of Preventive Medicine and Public Health* 46 (Suppl. 1): S50–S54.

Reardon, L., and S. Abdallah. 2013. Well-Being and Transport: Taking Stock and Looking Forward. *Transport Reviews* 33 (6): 634–657.

Rosenbloom, S. 2004. Mobility of the Elderly: Good News and Bad News. In *Transportation in an Aging Society: A Decade of Experience, Technical Papers and Reports from a Conference, November 7–9, 1999*. Bethesda, Maryland: Transportation Research Board.

Selwyn, N., S. Gorard, J. Furlong, and L. Madden. 2003. Older Adults' Use of Information and Communications Technology in Everyday Life. *Ageing & Society* 23 (05): 561–582.

Shergold, I., G. Lyons, and C. Hubers. 2015. Future Mobility in an Ageing Society: Where are We Heading? *Journal of Transport and Health* 2: 86–94.

Shergold, I., M. Wilson, and G. Parkhurst. 2016. The Mobility of Older People, and the Future Role of Connected Autonomous Vehicles. Project Report. Centre for Transport and Society, University of the West of England, Bristol, Bristol. http://eprints.uwe.ac.uk/31998. Accessed 31 October 2017.

Simonsick, E.M., J.M. Guralnik, S. Volpato, J. Balfour, and L.P. Fried. 2005. Just Get Out the Door! Importance of Walking Outside the Home for Maintaining Mobility: Findings from the Women's Health and Aging Study. *Journal of the American Geriatrics Society* 53 (2): 198–203.

Sixsmith, A. 2013. Technology and the Challenge of Aging. In *Technologies for Active Aging*, vol. 9, ed. A. Sixsmith and G. Gutman, 7–26. New York: Springer Science and Business Media.

Smith, A. 2014. Older Adults and Technology Use, Pew Research Center, Report 202.419.4500. http://www.pewinternet.org/2014/04/03/older-adults-and-technology-use/. Accessed 30 October 2017.

Sochor, J. 2013. The Reassuring Affects of ICT in Public Transportation: The Perspectives of Older Adults. *Gerontechnology.*

Sochor, J., and A. Nikitas. 2016. Vulnerable Users' Perceptions of Transport Technologies. In *Proceedings of the Institution of Civil Engineers*, Urban Design and Planning.

Souders, D.J., and N. Charness. 2014. Travel Safety and Technology Adoption by Elderly Populations.

Spinney, J.E., D.M. Scott, and K.B. Newbold. 2009. Transport Mobility Benefits and Quality of Life: A Time-Use Perspective of Elderly Canadians. *Transport Policy* 16 (1): 1–11.

Tacken, M., F. Marcellini, H. Mollenkopf, I. Ruoppila, and Z. Szeman. 2005. Use and Acceptance of New Technology by Older People. Findings of the International MOBILATE Survey: 'Enhancing Mobility in Later Life'. *Gerontechnology* 3 (3): 126–137.

TRACY Project. 2012. Work Package 2: Determining the State of the Art. Transportation Research Group (2006) 'Virtual mobility'. Available at http://www.trg.soton.ac.uk/vm/m-r.htm. Accessed 12 October 2006.

United Nations. 2015. World Population Ageing. http://www.un.org/en/development/desa/population/publications/pdf/ageing/WPA2015_Report.pdf. Accessed 31 October 2017.

Urry, J. 2004. The 'System' of Automobility. *Theory, Culture and Society* 21: 25–39.

World Health Organisation. 2002. *Active Ageing: A Policy Framework*. Geneva: Switzerland.

World Health Organisation. 2007. *Global Age-Friendly Cities: A Guide*. Geneva: Switzerland.

Zeitler, E., and L. Buys. 2015. Mobility and Out-of-Home Activities of Older People Living in Suburban Environments: 'Because I'm A Driver, I Don't Have A Problem'. *Ageing & Society* 35 (4): 785–808.

Ziegler, F., and T. Schwanen. 2011. 'I Like to Go Out to Be Energized by Different People': An Exploratory Analysis of Mobility and Wellbeing in Later Life. *Ageing & Society* 31 (5): 758.

Part II

Rural

4

Mobility, Transport and Older People's Well-Being in Sub-Saharan Africa: Review and Prospect

Gina Porter, Amleset Tewodros and Mark Gorman

Introduction

Mobility, or lack of it, is likely to be implicated in many facets of older people's lives. Research on this theme has been gathering pace rapidly over the last few years in response to demographic change associated with the ageing of populations (Schwanen and Paez 2010; Li et al. 2012). While a majority of work to date is focused on Western contexts,

G. Porter (✉)
Department of Anthropology, Durham University, Durham, UK
e-mail: r.e.porter@durham.ac.uk

A. Tewodros
HelpAge International Africa Region, Nairobi, Kenya
e-mail: amleset.tewodros@helpage.org

M. Gorman
HelpAge International, London, UK
e-mail: mgorman@helpage.org

© The Author(s) 2018
A. Curl and C. Musselwhite (eds.), *Geographies of Transport and Ageing*,
https://doi.org/10.1007/978-3-319-76360-6_4

the pace of population ageing is faster in developing than in developed countries and attention is thus now growing across the globe (Frye 2013), with recent work, for instance, in Taiwan (Liu and Tung 2014), China (Feng et al. 2013) and the Philippines (Petterson and Schmocker 2010), mostly with reference to urban contexts. In sub-Saharan Africa, where mobilities research focused on vulnerable user groups to date has concentrated principally on women's transport constraints, together with some work on children (mostly specifically on road safety), older people's mobility and transport needs are just beginning to attract attention. As in the Global North, older people now form a substantial proportion of the population: in many African contexts, however, their role in society has been made especially significant because of HIV and AIDS (Velkoff and Kowal 2006). Grandparents have been left supporting and caring for grandchildren, in the context of a significantly reduced or incapacitated middle generation resulting from parental deaths and ill health. The transport context also tends to be very different from that in richer countries, with very low private vehicle ownership, inadequate public transport systems, poor infrastructure and consequently much greater recourse, of necessity, to pedestrian travel. The implications for older people's well-being are considerable.

The first section of this chapter reviews the very limited research on older people's mobility and transport which has been conducted to date, with a particular focus on Anglophone countries in sub-Saharan Africa (notably Nigeria and South Africa). This is followed by reflections on recent mixed-methods research conducted through an NGO–academic collaboration with older people in rural Tanzania. Three themes given particular emphasis are, firstly, the particular significance of relationality in mobilities research with older people—other family and community members may substantially contribute to the shaping of older people's mobile lives; secondly, the importance of exploring potential new connectivities associated with increasing mobile phone and motorcycle-taxi usage among older people; and thirdly, the value of qualitative research, especially a co-investigation approach, to research with older community members. The final portion of the chapter is concerned with identification of significant research gaps where mobilities/transport-focused research with older people in sub-Saharan Africa is urgently needed.

Literature Review: Older People's Mobility and Transport in Africa

Studies of transport disadvantage and social exclusion in Africa are still relatively rare (Porter 2002; Lucas 2011); studies specifically addressing older people's mobility constraints and transport needs are particularly sparse. In terms of the spatial distribution of this research, urban-focused research to date appears to be limited to Nigeria (Odufuwa 2006; Ipingbemi 2010; Olawole and Aloba 2014) and rural research to a series of linked studies in Tanzania (Porter 2016; Porter et al. 2013a, b, 2014), discussed in the third main section of the paper. No studies specifically focused on older people's mobility and transport issues in urban areas have been identified for either eastern or southern Africa, except with respect to wider consideration of vulnerable populations (as, for example, Maart et al. 2007 on disabled people) and one countrywide transport review with a strong focus on older people. This is Venter's (2011) analysis of transport expenditure and affordability in South Africa drawing on data from the 2003 National Household Travel Survey.

Drawing on this literature, the following sub-sections review four key issues that affect older people's travel and transport: health and disability; transport availability and affordability; fear of harassment and crime; relationality, including the impact of caring responsibilities. In each component, we review the few studies specifically addressing older people's mobility constraints and transport needs and, additionally, consider some material beyond the transport/social exclusion domain that is relevant to that discussion.

Health and Disability Impacts on Mobility

Older people's mobility is often increasingly constrained by health/disability factors, especially as they move into their 70s and 80s. However, it would seem that in low- and middle-income countries, as in higher-income countries, active travel (walking and cycling) is beneficial to health and that this is often higher in poorer households than wealthier ones. A study by Laverty et al. (2015), drawing on data from a WHO report on Global Ageing and Adult Health (SAGE) for six middle-income countries (including Ghana and South Africa), found that there was wide variation in older adults (aged 50+) use of active travel for ≥150 min per week, with the lowest at 21% in South Africa and the highest at 58% in Ghana. Older people were less likely to use active travel for 150 min per week, as were women and those with higher levels of household wealth. They observe that high use of active travel was associated with reduced risks of being overweight, having a high waist-to-hip ratio, as well as a lower waist circumference, BMI and systolic blood pressure. Moderate use of active travel was associated with reduced risks of having a high waist-to-hip ratio and lower waist circumference and systolic blood pressure. This would suggest that although poorer older people usually have far less access to motorised transport than their wealthier counterparts, they may thereby gain health benefits.

On the negative side, however, the risk of falls while walking or cycling tends to be higher among older people than among youth and become a significant cause of morbidity and mortality with age. Moreover, the impact of falls can be to reduce mobility through the psychological consequences associated with fear of falling (Kalula et al. 2016). Unfortunately, there is little evidence, as yet, regarding risk factors for falls in the older population of sub-Saharan African countries; a study by Kalula et al. in Cape Town appears to be the first in South Africa (ibid.). Interestingly, people living in the predominantly black location studied reported fewer falls than those in white or mixed-ancestry neighbourhoods, which Kalula et al. suggest may relate to being engaged mostly in physically demanding occupations which give an advantage in terms of muscle reserve capacity and function and better maintenance of gait and balance. However, widespread hazards across much of Africa, such as poor infrastructure for pedestrians (uneven pavements, uncovered storm drains, etc.) and uneven roads and poorly regulated traffic for cyclists, may well increase danger levels for older people, especially those with poor eyesight, hearing or balance problems.

Despite potential constraints on older people's mobility, as they move into older age, the need for mobility to access health services tends to expand. In low- and middle-income countries in Africa, it is usually entirely incumbent on the patient to make their own way to the health centre, which can bring particular challenges for older people with poor physical strength and limited financial resources. A study of older people's access to health care in Uganda (Wandera et al. 2015), drawing on a nationally representative sample, thus suggested that four variables were significantly associated with healthcare access: age group, household poverty status, ownership of a bicycle as a means of transport and household major source of earnings. Older age (70+) and household poverty were associated with reduced access to healthcare in the last 30 days for older persons, while household ownership of a bicycle and earning of wages increased access to healthcare in the last 30 days. Issues of transport availability and affordability linked to the latter point are considered in the next section.

Transport Availability, Affordability and Ease of Use

Venter's (2011) analysis of transport expenditure and affordability in South Africa is particularly pertinent to this section. Drawing on data from the 2003 National Household Travel Survey, he focuses specifically on low income and mobility-constrained people, noting that approximately half of older people over 65 years ($N = 4522$) in that South African survey do not travel (i.e. did not leave their homes at all on the survey day), which is approximately twice that of the population as a whole. Analysis of transport problems (categorised into access problems, i.e. no available/suitable transport, affordability and service quality) by settlement type for older people throws further light on key issues. Venter shows that around one-third, or slightly more, older people in urban areas refer to access problems, as opposed to around two-thirds in rural areas; service quality is mentioned as a problem by around one-third of older adults in both rural and urban areas; affordability presents as a problem to 48% of poor rural older adults, but only 21% of poor urban older adults. It is interesting to note that the affordability issue is raised by a substantially higher proportion of rural poor older adults than the rural poor in general (36%), whereas in urban areas affordability figures for poor urban older adults are very similar to the poor urban population as a whole (23%). However, the biggest problem in rural areas for all groups is clearly access, with very similar figures (around two-thirds) for both older adults and the population at large, and with little variation according to wealth. Venter (p. 138) suggests that older adults who want to travel seem to be the worst off, because they often do not qualify for subsidies (currently mainly available to workers). He argues that *if there is a case to be made for expanding concessionary fares*, it could be made for older people, and this would add little to the cost of public transport subsidy. However, transport subsidies present a contentious issue in most country contexts, developing and developed.

Over the last decade, there have also been three Nigeria-focused transport studies based on urban household surveys with older people, all of which draw some attention to the inadequacy of current transport provision in terms of cost and availability, especially in contexts where older people must continue to travel for livelihood sustenance in the absence of social security. Odufuwa (2006), drawing on a large (3556) sample of people aged 60+ from household surveys in four Nigerian cities (including the former and current capital cities), presents a broad argument regarding urban older people's poor access to transport with particular reference to long waiting hours at bus stops, unfair charges, hostile behaviour of operators and inaccessible location of bus stops. Even so, he finds over 90% of respondents making use of public transport 1–2 times a week for medical- and work-related trips; this principally comprises taxis (with little usage of motorcycles or auto-rickshaws, which are considered too dangerous). Across the sites, the most important problems identified, in ranked order, were as follows: difficulty in boarding vehicles (31%), difficulty in alighting (28%), lengthy waiting time (15%), unfriendly attitude of operators and others (12%), poor/inaccessible bus stops (6%) and high fares (5%). The relatively low ranking of fare cost in this study is particularly intriguing (and not replicated in the two studies which follow): it may reflect the sampling procedure, for which no information is presented.

Ipingbemi's (2010) study of older people's mobility and travel characteristics draws on a smaller survey (290 people aged 60+) conducted across just one major south-west Nigerian city, Ibadan. As with the

Odufuwa's research, this study emphasises transport costs in the context of older people's need for continuing engagement in economic activities, in the absence of social security. Unsurprisingly, nearly 30% of their journeys were associated with livelihood activities (followed closely by health-related journeys). Key transport constraints observed varied between different city locations and, in addition to high transport fares, included reckless driving (especially by motorcycle operators), poor facilities (no shelter or seats), long waits at bus stops (sometimes due to drivers ignoring them and failing to stop unless there are other passengers) and the design of vehicles (steps too high) used for public transport. There was also reference to abuse by other passengers and by drivers (discussed further below).

Finally, Olawole and Aloba (2014) conducted a small household survey of 250 people aged 60+, sampled purposively in three residential zones of Osogbo, a state capital in south-west Nigeria. Interestingly, this survey puts traffic congestion as the greatest travel constraint for older people, but followed by other problems similar to those identified by Odufuwa and Ipingbemi (though again, as with Ipingbemi, with seemingly relatively higher ranking of transport cost issues than in Odufuwa's research).

Fear of Harassment and Crime

Older people (along with other age groups) can find journeys, whether involving walking alone or including public transport, daunting from a security perspective. However, data relating specifically to older age groups are very sparse.

In a rare example, Ipingbemi (2010) noted, in his study of older people in Ibadan, Nigeria, that there were reports of fear of abuse by other passengers and by drivers, including the impatience of transport operators dealing with them when they were alighting from the vehicle (especially when carrying loads), and pushing and shoving from other passengers in congested vehicles (for which see also Frye 2013: 18 on Jamaican cities).

Lloyd-Sherlock et al. (2016), using the same WHO (SAGE) data on global ageing and adult health for six middle-income countries as in Laverty et al. (2015), offer another perspective, in this case regarding pedestrian travel: the specific question on street safety included in the WHO survey is 'How safe do you feel when walking down your street alone after dark?'. These data reveal large national variations in reported crime fear, including between the two African countries in the survey, South Africa and Ghana; 65% of older South Africans felt unsafe on the street, compared to only 9% of older Ghanaians. However, in both countries, rates were much higher for older women than for men: among older women, street fear ranged from 10.1% in Ghana (compared to 7.4% men) to 68.3% in South Africa (versus 65.2 for men). In South Africa, being in a higher wealth quintile was associated with lower rates of fear (home and street), whereas in Ghana the opposite was found, with wealthier groups expressing higher rates of fear. Living in a rural area was associated with significantly lower rates of street fear in Ghana, but in South Africa rural location was associated with higher rather than lower levels of fear. The authors also note that the effect of sex on socialising outside the home varies by country: in Ghana female sex is associated with reduced socialising, in line with expectations based on gender relations in later life in the country.

The wider impacts on older people's lives can be substantial: in Lloyd-Sherlock et al.'s study, street fear was also found to be associated with less frequent use of outpatient health services, when controlling for self-reported health status, suggesting that older people with street fear are less inclined to leave the home to seek health care, even when they feel unwell. They suggest that it is likely that differences in reported rates of street fear are due to actual variations in fear of crime and violence: very high rates of fear and actual experience reported for South Africa are in line with the findings of other studies. However, they also emphasise that great care must be taken in attributing cause and effect in the relationship between fear and mobility, health and quality of life: while some limited insights about cause and effect can be obtained by looking at interactions between health, frailty and fear, their multivariate analysis suggests that although fear of crime may affect health and quality of life in some cases, the relationship is complex. A key point, however, is that frailty is '*by some distance*' (p. 1103) the most important determinant of fear, irrespective of other factors such as age or sex.

This same point will probably apply equally well with regard to travel on public transport.

Relationality, Including the Impact of Caring Responsibilities

This section emphasises the ways in which older people's mobility in Africa is socially produced, set within particular relations of power between themselves, their families and the wider community in which they live. The importance of relationality as a factor shaping African mobilities was flagged some years ago by Turner and Kwakye (1996), and see also Grieco et al. (1996) but has received little specific attention in subsequent transport/mobilities literature. Rare exceptions include Porter et al. (2013a), discussed below, Van Blerk (2016) on the relational im/mobilities of Ethiopian sex workers, and Porter et al. (2017: 15–17) on potential intergenerational tensions arising from relational im/mobilities from a child's viewpoint.

Turner and Kwakye's work in Accra drew specific attention to the way older people's mobilities in Africa often intersect with those of other family members. In that (and other West African) context(s), family fostering arrangements, whereby children are housed with older family members and expected to assist them, remain a widespread traditional practice.[1] They enable older grandmother traders to continue work by reducing their travel discomfort on Accra's overcrowded, uncomfortable and congested public transport system. Meanwhile, older people's reduced travel enables them to take the role of domestic anchors for larger households where many adults are engaged in activities across the city, since their trading then focuses on neighbourhood selling from the family home.

Another context in which older people's mobility may interlink closely with other family members is care-giving, whether to young (grand)children, where the parents have died or moved to another

[1] Though see Aboderin (2004) for reflections on decline in family support to older people in Accra associated with resource constraints and changing values.

location (as in Porter et al. 2013a for rural Tanzania, see below) or as carers of people with serious illness. With reference to the latter case, many older people are carers of people living with HIV and AIDS in resource-poor settings where there is limited access to anti-retroviral therapy. Chepngeno-Langat et al. (2011) emphasise the relatively neglected role of older male carers (50y+) in Nairobi slums and the significantly increased likelihood of disability and severe health problems among them, compared to male non-caregivers. The links between health-carer responsibilities and mobility have not, as yet, received attention in the transport/mobilities literature, but we can hypothesise that any reduction in older people's mobility outside the home, whether this reduces social or economic activities or both, is likely to bring a consequent reduction in well-being of the carer, thus fitting with the conclusions of Chepngeno-Langat et al. (2011) noted above. However, Ssengonzi (2009) to the contrary suggests that the care-giving responsibilities of those older people (especially women) in rural Uganda who have adult children affected by HIV and AIDS may require extensive travel to care for the sick. Clearly, much will depend on the personal/family circumstances of the carer concerned.

Building a Mixed-Methods Research Study Around Transport and Mobilities with Older People in Rural Tanzania

The significance of intergenerational intersectionality for older people's mobility in sub-Saharan Africa, reviewed above, is mirrored in our recent and ongoing research in rural Tanzania[2] (Porter et al. 2013a; Porter et al. 2014). This study commenced in 2012 in Kibaha district, Pwani Region, where we set out to explore older people's mobility and access to services and associated implications for health-seeking behaviour and livelihoods, following on from prior preliminary identification

[2]This research was conducted under the DFID-funded Africa Community Access Programme (AFCAP).

of transport unavailability as a significant problem for older people. In Tanzania, continuing access to livelihoods is frequently vital, not just for older people to support themselves, but now also to support young orphans and others in their care.[3] The mobility and mobility constraints older people face, which will impact strongly on their ability to act effectively in these—and other—roles, seemed to constitute a major knowledge gap.

As the (principally urban-focused) mobility literature reviewed above indicates, unlike in Western contexts, older people need to be mobile in order to access an income, so long as they are physically able. In rural areas, income from farming is frequently insecure, and multiplex livelihoods and off-farm income often appear to offer the best route to survival (e.g. Bryceson 2002; Gladwin et al. 2001). However, this very often requires travel to the nearest market or service location, causing particular difficulties, for instance, for older women traders. Ill health and infirmity may introduce further problems for older people, in a rural walking world where pedestrian transport dominates among all age groups. Reduced pedestrian mobility, due to infirmity and the unaffordable cost or unavailability of motorised transport, may help to limit older people's access to work and vital health care, thus reinforcing their poverty: thus a vicious circle develops in which mobility restrictions form a key component.

Identifying Hypotheses Around Rural Older People's Mobility for Field Testing

In the absence of detailed information on the rural transport and mobility issues faced by older people, a series of hypotheses were put forward initially for examination in our Kibaha study. These drew on the very limited (and mostly urban-focused) published literature available at that time, but also on our wider understandings and experiences in rural Africa, on HelpAge's extensive experience of working with older people

[3]Approximately 20% of 3000 child respondents surveyed in 2007/2008 in a child mobility study lived with people other than their parents—in South Africa, Malawi and Ghana, respectively, 14, 9 and 9% live with grandparents (usually grandmother alone); the remainder lived with other relatives/foster parents, many of whom are older people (Porter et al. 2010).

worldwide and through preliminary discussions with older people living in the study district itself.

Health and Disability Impacts on Mobility in Rural Areas

HelpAge's previous work in Kibaha district had emphasised that, as people move into older age, the need for mobility to access health services expands, yet they are hampered by transport-related factors: this was a key reason why the agency wished to explore mobility issues.

Older travellers may face difficulties around specific health problems sometimes associated with old age, such as urinary incontinence among women due to earlier obstetric problems (e.g. obstetric fistula and related conditions). This had not been reported in the urban mobilities literature on older people, but women with such conditions are more likely to be encountered in remoter rural areas where access to obstetric services is poor. Older people are also likely to be at disproportionate risk of road traffic accidents (a major cause of injury and death across Africa) because of age-related physical and cognitive changes.

Another potential health issue identified in other rural locations has been the demands of load carrying on women from childhood and onwards, which *appear* to impact severely on health and quality of life as they enter and experience old age (though we are unaware of any evidence base to support this hypothesis; see Porter et al. 2013b). The implications of Africa's transport gap and consequent dependence on pedestrian head-loading (often designated a female activity) have received remarkably little attention. The particular plight of older women in accessing fuelwood, water and markets needed further investigation.

Transport Availability, Affordability and Ease of Use in Rural Contexts

We hypothesised that, as in urban areas, lack of reliable low-cost transport and restricted mobility was likely to severely affect older people's access to clinics, pension points (where pensions are provided), paid work, livelihood opportunities, churches, mosques and other faith institutions, participation in social networks, and other facilities and services important to their lives, with negative impacts on their health and well-being. Long walks to access a transport route or to services were likely to present a serious hurdle, particularly to less fit older people and people with a disability, and especially along rural routes crossing difficult terrain, and in the rains. Even where regular transport is available, low incomes and poverty could well limit access: older people, especially women carers, often appear to be among the poorest in communities, thus probably those least able to afford transport fares. Even when older people are able to access public transport, they may face numerous difficulties (as Ipingbemi 2010 suggested in his Ibadan urban study).

Fear of Harassment and Crime in Rural Contexts

Our research pre-dated the availability of publications by Ipingbemi (2010) and Lloyd-Sherlock et al. (2016), but prior work by the lead author with children suggested that some difficulties that older people face in rural areas were likely to be similar to those observed in youth mobility contexts: harassment by transport operators, being cheated on fares by operators (Porter et al. 2010).

Relationality, Including the Impact of Caring Responsibilities, in Rural Contexts

As noted above, older people's mobility in Africa is set within particular relations of power between themselves, their families and the wider community in which they live. We can expect considerable diversity of experience and access to power amongst older people, according to age, gender, ethnicity, socio-economic status, family composition (dependants), occupational history, infirmity/health, personal mobility status, density of service provision, etc. It is important to assess how this diversity impacts on transport usage, suppressed journeys, mobility and access to services and other elements important to older people's well-being. Thus, we hypothesised that very old and infirm people, in particular, may face a lack of power and access to wider decision-making processes (similar to that experienced by children). Where this is the case, their views are less likely to be heard and their transport and mobility needs even less likely to be met than those of other groups. Consequently, we needed to put a particular focus on working with older people in novel ways to ensure their voices were heard.

In terms of caring responsibilities, older people's role as carers of young children whose parents were dead or working in town was particularly striking in rural Kibaha (as in much of rural Tanzania). This led us to hypothesise the likely interconnectedness of child-adult mobilities, with possible negative and positive impacts. Negative impacts could include a reduction in the educational, health and livelihood opportunities of children and young people in their care, thus reducing overall long-term potential for poverty eradication. For instance, mobility and access constraints would be likely to impact strongly on older people's ability to earn income, with consequent impact on their ability to feed, clothe and educate children in their care. HelpAge International has argued that access to livelihoods had been inadequately considered in an older people's context (they are often treated by government, academics and others as if they are outside the working population). On the other hand, older people might gain access to services not only directly but also indirectly through both adults and children in the community. The relationality between children and older people's lives has been considered in general terms (e.g. Whyte et al. 2004), but needed analysis in a mobility context (as in Turner and Kwakye 1996). Thus, impacts on older people of other household and community members' mobility need to be considered, especially regarding migration, which may affect indirect access to services via family helpers.

Exploring the Potential to Improve Rural Older People's Well-Being Through Improved Connectivities

HelpAge International's institutional mandate encouraged us to explore potential routes to improving older men and women's access to services in rural Kibaha. These were likely to vary from those open to younger people in their communities and from urban settings. Bicycle usage, for instance, may be impossible for older women who have never had time/ opportunity to learn to cycle. Older people with disabilities are particularly disadvantaged, such that even mobile service provision to settlement centres may not serve them adequately.

The crucial importance of connectedness to family for older people in Africa needs to be set within the context of limited work potential, ill health and lack of income security: social bonds are likely to be essential to securing care and financial support in old age. In many African societies, giving money is a way younger kin traditionally pay respect and show affection and care for older relatives, but when the younger generation has migrated elsewhere, it may be difficult for older people to achieve the sustained interaction necessary to maintain such links. Mobility and access to affordable transport are likely to be key factors in sustaining social networks; however, early observation in Kibaha suggested that mobile phones could also play a growing role. The potential for mobile phone use (expanding dramatically across Africa) to substitute virtual for physical mobility to the advantage of older people looked considerable: current and potential uses among older people needed investigation.

The Kibaha Research Project

We selected 10 settlements with varying accessibility to roads and health services for detailed research to investigate the hypotheses noted above, utilising a three-strand research methodology:

1. Community co-investigation with a small group of 12 older people (8 men, 4 women, all aged between 59 and 69), to establish key issues for further investigation and analysis, with planned working in one focus settlement after a one-week training workshop, $N = 74$ transcripts. This first strand of the research was particularly novel.
2. Academic-led qualitative studies, $N = 194$ in-depth interviews using checklists, conducted with older people and other key informants (health workers, transport operators, settlement leaders).

3. Finally, a survey questionnaire to older people, $N = 339$, obtained by selecting households along settlement transects, interviewing those older people aged 60+ in each household and aiming for a minimum of 30 completed questionnaires per settlement.

Activities 2 and 3 were both based on findings and key questions identified from activity 1, but in this case extending to all 10 settlements. Qualitative interview data from the older people's work and the academic-led qualitative interviews were analysed thematically, and information for key themes was triangulated with findings from the SPSS survey data analysis. Full details of the research process are available in Porter et al. (2014).

A further note may be helpful regarding the co-production component of the study. The key difference to note between co-production of knowledge and co-investigation is that co-investigation represents a further move along a continuum of engagement with less powerful participants. While participation and dialogue characterise co-production of knowledge, in co-*investigation* such partners are actively engaged in the research process, as peer researchers (Porter 2016). Some brief reflections on the value of taking a co-investigation approach to research with older community members may be useful here, since we are aware of only one earlier study where co-investigation has involved the recruitment of older people (a HelpAge study by Ibralieva and Mikkonen-Jeanneret 2009). Firstly, our older people peer researchers seem to have enjoyed the research process greatly, and while this may have been partly a reflection of the small payments they received for their work, it is also the case that they were sufficiently engaged in the study to independently decide to continue and work across all ten study villages after their allocated work component was complete. Secondly, as noted above, their findings were crucial to the design of the academic-led components of the study. Subsequently, at the national workshop in Dar es Salaam which followed, where the Kibaha project findings were presented to government and NGO staff, the older people's team participated with great enthusiasm. The success of the project and the co-investigation approach was sufficient to encourage HelpAge Tanzania to undertake further studies, including one focused on intergenerational relations, with both children and older people trained as peer researchers (Mulongo et al. 2014) and to then follow up with further work on older people's mobility in another region of Tanzania.

Our main transport findings and links back to the initial hypotheses are reported in full in Porter et al. (2013a). Of these, the new connectivities we discovered associated with increasing mobile phone and motorcycle-taxi usage among older people are likely to be of particular significance for the future in rural areas of Africa like Kibaha. Motorcycle taxis (boda-boda) are now the main modes of transport in this district, except along the paved road. According to local inhabitants, they have spread rapidly in all the Kibaha study settlements since c. 2007–2009, before which transport was limited mainly to bicycle taxis. Their uptake has been facilitated by the availability of cheap imported Chinese motorcycles and the fact that they offer a major employment opportunity for rural youth. We concluded that boda-boda have transformed rural lives in this district, even where older people are concerned: 18% of older women and 31% of older men had used one in the week before the survey ($n = 339$)—they are now ubiquitous. While older people would *prefer* other motorised transport (i.e. buses or minibuses which are cheaper, much safer and more comfortable than boda-boda), the only real alternative is usually walking. They are, of course, especially important in health emergencies, but they are also important for livelihoods. The small scale of farming among many older people makes boda-boda feasible for transport of farm produce and farm inputs. We found some older people taking goods to market by boda-boda and carrying farm inputs home from town by the same means. Moreover, some urban traders were coming into the district by boda-boda to purchase at their farms. Boda-boda were even important in our one study settlement located on the paved road, because though other vehicles are available here, boda-boda are able to take people to their doorstep, and continue to run at night when other transport services have stopped.

Meanwhile, mobile phones are a complementary new connector in Kibaha (Porter 2016). Cell phone expansion has been remarkable in Tanzania as across much of Africa, despite user challenges (airtime cost, network, charging). Airtime costs have led to various adapted (low-cost) modes of use—'buzzing' (i.e. cutting off a call before a recipient answers so they know they should call back), SMS (short messaging), etc. Our Kibaha survey of older people found that mobile phones were owned by 41% of the older men surveyed and 15% of older women. They had often been acquired as a gift from their children living in town. However, phones were also widely available to older people through relatives and friends: sharing of phones was extremely common, such that almost all—both women and men—had access when they needed one. The implications for transport users are considerable, since they can now call transport operators to come to pick them up, which greatly improves access to transport. Transport operators, meanwhile, including boda-boda drivers, find a mobile phone increasingly essential to business (in an increasingly competitive business environment). Boda-boda services are, of course, especially efficient when ordered by mobile phone. Some older people reported having as many as five boda-boda operator numbers on their mobile phone. Thus, mobile phones are now widely used to organise boda-boda transport both in emergencies and every day. Phones have also, however, led to some reported reduction in travel among older people: *I don't have to travel so much nowadays—maybe when there is a funeral or a crucial thing for me to travel, but for minor things I use my brother's phone and we talk* (Woman 66y); *Nowadays I don't travel much to go to my children in town, instead we talk (on the phone) and solve our problems where possible* (Woman 78y).

Mobile phones were also being used to transfer money, such that, in some cases, children in town were now sending money to their older parents instead of bringing it to the village. This has had the benefit of reducing the time and cost incurred in travel and the not inconsiderable risks of a road accident or theft: *I use M-PESA; my children usually send money through my chip (Vodacom-number), then they call my friend through his phone telling how much they have sent through my Vodacom-line, so I just go with my chip to the Vodacom shop to take money* (Man 66y).

There is much evidence, overall, of improved well-being as a result of these phone and boda-boda innovations. In particular, they have brought a substantial improvement in access in remote areas, not least for older people, especially regarding emergency health travel. Moreover, solar phone charging, airtime sales and boda-boda ownership/rental are now part of the livelihood repertoire of some village elites (including some older people).

The mobile phone certainly supports social interaction in stretched households (where children are in town and grandchildren left with grandparents) and can facilitate remittances, while reported reductions in travel *overall* due to the increasing use of phones—especially benefits remote populations and the infirm. At the same time, of course, there is also a downside if—as some older people observed—they then see very little of family members: *Most older people have phones now. They call their children who are far away. If you don't remind the children they forget you and your needs.* (Man 71y, caring for 5 young orphaned grandchildren). It is possible that reduced face-to-face interaction may eventually leave some older people feeling lonely and isolated.

Key Research Gaps in Mobilities/Transport-Focused Work with Older People in Africa

From the literature review, it is evident that we have not, as yet, progressed far—or deeply—in our understanding of older people's mobility issues across sub-Saharan Africa. This research gap needs urgent attention, given the fact that both absolute numbers and the share of older people are on the increase in Africa, as elsewhere (Schwanen and Paez 2010). The impact of older people's (limited) access to transport and mobility on their livelihood opportunities, health and well-being has particular significance in African contexts where social security is lacking and health services particularly inadequate.

In terms of *areal coverage* across Africa, apart from Venter's South Africa countrywide analysis, there are only a few studies of older people's mobility in urban contexts (West Africa), just the one rural study in Tanzania and occasional passing reference to the potential significance of older people's mobility and access to transport in research focused on other vulnerable groups. Conditions in most of the continent remain essentially unexplored. It is important not to assume that older people's constraints and transport needs are ubiquitous: variations may occur according to topography, population density, climatic conditions, cultural context, etc., but, as yet, we have little indication what patterns these may take.

Age and gender disaggregation is important in all mobilities research, but is often inadequately considered. Age disaggregation within the 'older people' category is clearly vital, since physical mobility tends to decline over time with physiological changes; the mobility characteristics and transport needs of older people over c. 75 years may be quite different from people a decade younger. Additional *disabilities data* will also be extremely important. *Gender*, meanwhile, is likely to continue to shape desired and actual activity patterns and associated mobilities (albeit depending, in part, on cultural context). It is important to note that many older African women have far greater freedom to travel than their pre-menopausal counterparts, so long as they have access to resources and the physical capacity to do so (Porter 2011). However, gender is often a key determinant of financial resources, with older women commonly far less able to afford means of transport or even transport fares than older men.

It is also essential to examine older people's mobilities within the wider context of the *household composition and intergenerational relations*, and the associated mobilities in which they are enmeshed. The significance of relationality in mobility contexts was strongly emphasised in our Tanzania study, where older people's livelihood contributions were found to be crucial to many young people in their care, and where young people also often assisted older people with work like carrying firewood and water from locations beyond the homestead, or travelling to the nearest market town to buy vital medicines. Some of the most disadvantaged older people in our Tanzania study were those residing in single-occupancy households[4] without family living nearby.

Reflections on *mobility changes along the life course*, meanwhile, may draw our attention to the complexity of changing bodily capacities (see Ingold and Vergunst 2008: 17) and their potential wider impacts. Younger people travelling *with* older people may perhaps change the tenor of an older person's journey—possibly on the one hand bringing

[4] 12.1% of women and 10.6% of men lived in single-occupancy households.

companionability and security and on the other hand the stress of keeping pace with a younger, more able, energetic walker. It is important to think of mobility not just in terms of getting from A to B but also in terms of emotion and *affect*. Research on mobilities in Africa has barely started in this field but has the potential to contribute to a much deeper understanding of mobile lives.

The *intersection of virtual with physical mobility* is a rapidly emerging issue, as observed in the Tanzanian context. For older people, there can be potentially enormous advantages in terms of feasible reductions in both financial and physical costs associated with reduced travel. M-Health services are also currently expanding very rapidly in many African countries. Transport researchers are only slowly beginning to recognise the importance of including consideration of virtual mobility in planning for transport and access to services: more joined-up thinking in this arena is essential. However, there are also potential disadvantages, particularly for older people, associated with the decline of face-to-face interactions which may ensue wherever virtual mobility expands—this will need careful monitoring, especially if it is accompanied by contraction in rural service provision. Loneliness and isolation can result when phone contact curtails co-presence, as research in other contexts is beginning to suggest (Hardhill and Olphert 2012).

Insofar as *research methodologies* employed are concerned, we submit that there is a need for more mixed-method studies which will extend survey research of the kind which has been conducted recently in Nigeria by academics working in transport studies into a more grounded analysis. That research has been valuable, in itself, in drawing attention to older people's mobility needs, not least because the numbers produced in surveys are generally deemed by the transport sector an essential pre-requisite for policy and planning. However, mixed-methods approaches are likely to bring to the fore a much stronger view of issues from the grassroots, especially if the starting point is community co-investigation involving older people themselves. When this is then followed up by further qualitative research with key informants such as transport operators, the questions subsequently included in large-scale surveys are more likely to be precisely pertinent to understanding and resolving issues on the ground.

Action research studies, whereby interventions are made in communities and then carefully monitored would be particularly valuable. Information from our Tanzania study suggests this might include Intermediate Means of Transport (IMT) interventions such as specially designed light carts for transporting water and firewood (a major burden in intra-village/neighbourhood transport for many older people), and development and trials of a safety harness for use on motorcycle taxis, or other means whereby boda-boda might be adapted to make it safer and more comfortable to older people and also to examine feasible alternatives, such as a more advanced ambulance trailer/tricycle, especially in the context of travel of sick older people to health centres. More broadly, piloting of a community-run emergency health service with a suitable 3- or 4-wheel vehicle which is easy to board and comfortable, to carry older people and other vulnerable groups such as people with disabilities to the nearest district health facility, with a small fund to provide fares for emergency treatment, could be valuable in some contexts. Community '*Transport to health*' clubs (similar to funeral clubs), where small regular contributions are made by individuals and/or national social assistance funds towards emergency hospital transport, to help vulnerable groups prepare for health emergency expenditures, might well form a useful linked intervention, together with community arrangements to support improved emergency night transport in remoter settlements, based on designated community cell phone links with a small number of local private transport operators. In promoting trials of such interventions, it will be important to emphasise the broader benefits of paying attention to older people's transport needs, given competing demands on constrained resources. As noted above, the mobility of different age groups is complexly intertwined: interventions to assist older people can contribute substantially to wider community development.

Concluding Reflections

Clearly, the research task in view is enormous. Our experience suggests that the best prospect for building a sound evidence base is in collaborative research involving not just academic researchers and NGOs but also the older people directly affected by transport and mobility constraints. However, while high-quality research is essential to understanding these issues, it is essential to bear in mind that any required interventions are likely to depend, at least in part, on bringing the transport sector—and other sectors such as health—directly into the research nexus. This ideally needs to happen from the start of the research. Transforming evidence into policy and practice is particularly challenging in the transport sector which is dominated by male, middle-aged, middle-class engineers whose principal focus is road construction rather than transport services and where there is still a common reluctance to engage with users or with qualitative data (as has been observed elsewhere with reference to gender issues, see Porter 2008). Olawole and Aloba (2014) observe the total absence of national transport policy on the travel and mobility of older people in Nigeria: the same can probably be said for most African countries (but would be worthy of further investigation).

There is urgent need for government action where older people's mobility is concerned. We have identified a few champions and allies in the transport and health sectors in the course of our Tanzanian research, principally by virtue of our fortuitous inclusion in a wider community access research programme (AFCAP), which is focused principally on road engineering but also includes transport services research in diverse areas, including maternal health. Extending linkages and building on alliances with researchers, policy makers and practitioners in this way, are likely to be crucial in efforts to improve transport and mobility for older people in the longer term.

Acknowledgements Funding: The Africa Community Access Programme [AFCAP] funded our field research in Tanzania.

References

Aboderin, I. 2004. Decline in Family Material Support for Older People in Urban Ghana. *Journal of Gerontology, Series B* 59 (3): S128–S137.

Bryceson, D.F. 2002. The Scramble in Africa: Reorienting Rural Livelihoods. *World Development* 30 (5): 725–739.

Chepngeno-Langat, G., N. Madise, M. Evandrou, and J. Falkingham. 2011. Gender Differentials on the Health Consequences of Care-Giving to People with AIDS-Related Illness Among Older Informal Carers in Two Slums in Nairobi, Kenya. *AIDS Care* 23 (12): 1586–1594.

Feng, J., M. Dijst, B. Wissink, and J. Prillwitz. 2013. The Impacts of Household Structure on the Travel Behaviour of Seniors and Young Parents in China. *Journal of Transport Geography* 30: 117–126.

Frye, A. 2013. Disabled and Older Persons and Sustainable Urban Mobility. Thematic Study Prepared for Sustainable Urban Mobility: Global Report on Human Settlements 2. http://www.unhabitat.org/grhs/2013.

Gladwin, C.H., A.M. Thomson, J.S. Peterson, and A.S. Anderson. 2001. Addressing Food Security in Africa via Multiple Livelihood Strategies of Women Farmers. *Food Policy* 26 (2): 177–207.

Grieco, M., N. Apt, and J. Turner. 1996. *At Christmas and on Rainy Days: Transport, Travel and the Female Traders of Accra*. Aldershot: Ashgate.

Hardill, I., and C.W. Olphert. 2012. Staying Connected: Exploring Cell Phone Use Amongst Older Adults in the UK. *Geoforum* 43: 1306–1312.

Ibralieva, K., and E. Mikkonen-Jeanneret. 2009, July. *Constant Crisis: Perceptions of Vulnerability and Social Protection in the Kyrgyz Republic*. HelpAge International.

Ingold, T., and J.L. Vergunst. (2008). Introduction. In *Ways of Walking: Ethnography and Practice on Foot*, ed. T. Ingold and J.L. Vergunst. Aldershot: Ashgate.

Ipingbemi, O. 2010. Travel Characteristics and Mobility Constraints of the Elderly in Ibadan Nigeria. *Journal of Transport Geography* 18 (2): 285–291.

Kalula, S., M. Ferreira, G. Swingler, and M. Badri. 2016. Risk Factors for Falls in Older Adults in a South African Urban Community. *BMC Geriatrics* 16 (51): 1–11.

Laverty, A., R. Palladino, J. Lee, and C. Millett. 2015. Associations Between Active Travel and Weight, Blood Pressure and Diabetes in Six Middle Income Countries: A Cross-Sectional Study in Older Adults. *International Journal of Behavioral Nutrition and Physical Activity* 12 (65): 1–11.

Li, H., R. Raeside, T. Chen, and R.W. McQuaid. 2012. Population Ageing, Gender and the Transportation System. *Research in Transportation Economics* 34: 39–47.

Liu, Y.-C., and Y.-C. Tung. 2014. Risk Analysis of Pedestrians' Road Crossing Decisions: Effects of Age, Time Gap, Time of Day and Vehicle Speed. *Safety Science* 63: 77–82.

Lloyd-Sherlock, P., S. Agrawal, and N. Minicuci. 2016. Fear of Crime and Older People in Low- and Middle-Income Countries. *Ageing & Society* 36: 1083–1108.

Lucas, K. 2011. Making the Connections Between Transport Disadvantage and the Social Exclusion of Low Income Populations in the Tshwane Region of South Africa. *Journal of Transport Geography* 19 (6): 1320–1334.

Maart, S., A.H. Eide, J. Jelsma, M.E. Loeb, and M. Ka Toni. 2007. Environmental Barriers Experienced by Urban and Rural Disabled People in South Africa. *Disability & Society* 22 (4): 357–369.

Mulongo, G. with A. Tewodros, L. Ndamgoba, A. Heslop, N. Idehen, and J. Milovanovic. 2014. Study Report on Extending Healthy Ageing Through the Life Course: Intergenerational Interventions in Rural Tanzania. Unpublished Paper, HelpAge Tanzania, September.

Odufuwa, O.B. 2006. Enhancing Mobility of the Elderly in Sub-Saharan Africa Cities Through Improved Public Transportation. *IATSS Research* 30 (1): 60–66.

Olawole, M.O., and O. Aloba. 2014. Mobility Characteristics of the Elderly and Their Associated Level of Satisfaction with Transport Services in Osogbo, Southwestern Nigeria. *Transport Policy* 35: 105–116.

Pettersson, P., and J.-D. Schmocker. 2010. Active Ageing in Developing Countries?—Trip Generation and Tour Complexity of Older People in Metro Manila. *Journal of Transport Geography* 18 (5): 613–623.

Porter, G. 2002. Living in a Walking World: Rural Mobility and Social Equity Issues in Sub-Saharan Africa. *World Development* 30 (2): 285–300.

Porter, G. 2008. Transport Planning in Sub-Saharan Africa. Progress Report 2. Putting Gender into Mobility and Transport Planning in Africa. *Progress in Development Studies* 8 (3): 281–289.

Porter, G. 2011. 'I Think a Woman Who Travels a Lot Is Befriending Other Men and That's Why She Travels': Mobility Constraints and Their Implications for Rural Women and Girl Children in Sub-Saharan Africa. *Gender, Place and Culture* 18 (1): 65–81.

Porter, G. 2016. Mobilities in Rural Africa: New Connections, New Challenges. *Annals of the American Association of Geographers* 106 (2): 434–441.

Porter, G., A. Heslop, F. Bifandimu, E. Sibale, A. Tewodros, and M. Gorman. 2014. Exploring Intergenerationaliy and Ageing in Rural Kibaha Tanzania: Methodological Innovation Through Co-investigation with Older People.

In *Intergenerational Space*, ed. R. Vanderbeck and N. Worth. London: Routledge.

Porter, G. with K. Hampshire, A. Abane, A. Munthali, E. Robson, and M. Mashiri. 2017. *Young People's Daily Mobilities in Sub-Saharan Africa. Moving Young Lives.* London: Palgrave Macmillan.

Porter, G., K. Hampshire, A. Abane, E. Robson, A. Munthali, M. Mashiri, and Augustine Tanle. 2010. Moving Young Lives: Mobility, Immobility and Urban 'Youthscapes' in Sub-Saharan Africa. *Geoforum* 41: 796–804.

Porter, G., A. Tewodros, F. Bifandimu, M. Gorman, A. Heslop, E. Sibale, A. Awadh, and L. Kiswaga. 2013a. Transport and Mobility Constraints in an Aging Population: Health and Livelihood Implications in Rural Tanzania. *Journal of Transport Geography* 30: 161–169.

Porter, G., K. Hampshire, C. Dunn, R. Hall, M. Levesley, K. Burton, S. Robson, A. Abane, M. Blell, and J. Panther. 2013b. Health Impacts of Pedestrian Head-Loading: A Review of the Evidence with Particular Reference to Women and Children in Sub-Saharan Africa. *Social Science and Medicine* 88: 90–97.

Schwanen, T., and A. Paez. 2010. The Mobility of Older People—An Introduction. *Journal of Transport Geography* 18 (5): 591–668.

Ssengonzi, R. 2009. The Impact of HIV/AIDS on the Living Arrangements and Well-Being of Elderly Caregivers in Rural Uganda. *AIDS Care* 21 (3): 309–314.

Turner, J., and E. Kwakye. 1996. Transport and Survival Strategies in a Developing Economy: Case Study Evidence from Accra, Ghana. *Journal of Transport Geography* 4 (3): 161–168.

Van Blerk, L. 2016. Livelihoods as Relational Im/Mobilities: Exploring the Everyday Practices of Young Female Sex Workers in Ethiopia. *Annals of the American Association of Geographers* 106 (2): 413–421.

Velkoff, V.A., and P.R. Kowal. 2006. Ageing in Sub-Saharan Africa: The Changing Demography of the Region. In *Ageing in Sub-Saharan Africa: Recommendations for Furthering Research*, ed. B. Cohen and J. Menken, 55–81. Washington, DC: National Academy of Sciences.

Venter, C. 2011. Transport Expenditure and Affordability: The Cost of Being Mobile. *Development Southern Africa* 28 (1): 121–140.

Wandera, S.O., B. Kwagala, and J. Ntozi. 2015. Determinants of Access to Healthcare by Older Persons in Uganda: A Cross-Sectional Study. *International Journal for Equity in Health* 14 (26): 1–10.

Whyte, S.R., E. Alber, and P.W. Geissler. 2004. Lifetimes Intertwined: African Grandparents and Grandchildren. *Africa* 74 (1): 1–5.

5

A *Window to the Outside World*. Digital Technology to Stimulate Imaginative Mobility for Housebound Older Adults in Rural Areas

Gillian Dowds, Margaret Currie, Lorna Philip and Judith Masthoff

Introduction

This chapter presents findings from a study that explored the potential for digital technologies to enhance the well-being of largely housebound,[1] older adults in rural areas. Current demographic ageing processes mean that the absolute numbers of older people are increasing worldwide. From 2025 to 2050, the older population is projected to almost double to 1.6 billion globally, whereas the total population will grow by just 34% over the same period (United States Census

[1]'Largely housebound' encompasses older adults who self-identified as being 'less able to get out and about as they used to, due to a long-term physical health condition'.

G. Dowds (✉) · L. Philip
School of Geosciences, University of Aberdeen,
Aberdeen, UK
e-mail: Gillian.dowds1@abdn.ac.uk

L. Philip
e-mail: L.Philip@abdn.ac.uk

© The Author(s) 2018

A. Curl and C. Musselwhite (eds.), *Geographies of Transport and Ageing*,
https://doi.org/10.1007/978-3-319-76360-6_5

Bureau 2016). With increasing age comes the likelihood of living with at least one long-term illness, which can compromise personal mobility (AgeUK 2015). Many older adults live healthy, active and independent lives, but sizable proportions globally, live with one or more long-term conditions. Chronic, long-term conditions already account for more than 87% of lost 'healthy' years amongst over 60s in low-, middle- and high-income countries and this percentage is expected to keep increasing (World Health Organization 2011: 9–10). Housebound, or largely housebound adults can find it difficult to 'get out and about', to interact with other people and to engage with the natural environment, all activities known to promote well-being (Nordbakke and Schwanen 2014; Ziegler and Schwanen 2011; Spinney et al. 2009). The ability to remain engaged in these activities can be particularly challenging for older adults in rural areas. The provision of transport is generally prioritised for 'utility' trips, or those regarded as essential over those associated with leisure or general enjoyment (Parkhurst et al. 2014). For those in rural areas who are unable to access public transport, engagement in such activities associated with well-being can be further prohibited, due to declining services in rural areas diminishing the presence of 'community hub' spaces to meet others and the likelihood of family members being dispersed rather than living locally. Thus, challenges with accessibility due to an absence of places to socialise with others in rural areas can further inhibit opportunities to socialise with others in one's community. The importance of maintaining participation in activities associated with well-being, including social activities and involvement with the natural environment, is supported by international 'Active Ageing'

M. Currie
Social, Economic and Geographical Sciences, The James Hutton Institute, Aberdeen, UK
e-mail: margaret.currie@hutton.ac.uk

J. Masthoff
School of Natural and Computing Science, University of Aberdeen, Aberdeen, UK
e-mail: J.Masthoff@abdn.ac.uk

strategies, which aim to promote health, participation and security, to enhance quality of life in later life (e.g. World Health Organization 2002; International Council on Active Ageing 2001–2016). Digital technologies provide new opportunities for maintaining existing and developing new forms of social interactions (e.g. Damant and Knapp 2015) and provide a new means for older adults, especially those who are largely housebound, to remain engaged with activities that they are no longer able to attend or participate in themselves.

The study adopted a two-phase design whose overall aim was to inform the development of a technology that could enhance the sense of engagement older adults in rural areas had with their local community and to examine the impacts of whether use of such a technology could potentially promote well-being. The chapter is structured as follows. Firstly, we contextualise the study, reviewing how health and non-health-related barriers combine to create challenges for older rural adults' involvement in local activities that could potentially be addressed, at least in part, by digital technology applications. Next, the methods adopted in the study are described, followed by an overview of the findings from this research, including a detailed description of the technology concept that was designed and then evaluated. Finally, the theme of 'engagement in meaningful activities', articulated by many participants in the study, is appraised with particular reference to *imaginative mobility*. The potential for digital technologies such as that designed for this research to facilitate increased opportunities for this will be deliberated.

Research Context

Demographic ageing is an international phenomenon (National Institute on Aging 2011), and promoting and sustaining the well-being of older adults are a priority for many national governments. Worldwide, most countries are experiencing demographic ageing, a result of declining fertility rates and increasing life expectancy (World Health Organization 2011). Globally, the highest proportion of older people is found in Europe and it is predicted that it will continue to have the oldest population profile until at least 2050, by which stage the proportion of the European population aged 65 or over is estimated to be 37% (United Nations 2001: 12). Within world regions and individual countries, the spatial patterns of demographic ageing are not uniform. In Western nations, demographic ageing is most pronounced in rural areas. In the USA, for example, the proportion of those aged 65 and over in 2006 was higher in rural (15%) than in urban (12%) counties (Philip et al. 2012). An urban–rural difference is replicated across much of Europe, including the UK, where demographic ageing is particularly pronounced in remote rural areas, a geographic pattern predicted to continue for the foreseeable future (Blake 2009). This pattern is largely due to the in-migration of pre-retirement age residents, and older adults ageing, and dying, in place (Philip et al. 2012).

The likelihood of developing one or more chronic, non-communicable conditions such as heart disease, chronic obstructive pulmonary disease and stroke, which combined account for two-thirds of deaths worldwide (Lozano et al. 2012) increases with age. The Scottish Government (2013, no page numbers) noted that '27% of people aged 75–84 have two or more [long-term conditions]'. Many of these people live in their own home; some receive regular care and support from health and social care professionals. As demographic ageing becomes more pronounced, the absolute number of older adults requiring health and/or social care interventions increases, with implications for the design and delivery of health and social care services. Measures to help older people manage their long-term condition at home can reduce hospital admissions, in a process the Health Service in Scotland calls 'Enablement' (The Scottish Government 2010). Likewise, more general measures to promote Enablement and Active Ageing, and to enhance the overall well-being of older adults with chronic, long-term conditions, may also reduce the likelihood of inpatient care being required, thus making savings to the health service. The promotion of opportunities for social interaction and engagement with the outside world is particularly important for supporting the well-being of older adults whose chronic condition(s) creates mobility challenges which limit their ability to 'get out and about' unassisted, rendering them largely housebound and socially isolated. In the Global North, the increasing likelihood of living with one or more chronic health conditions during later life can challenge personal preferences for independent living at home as the preferred option as opposed to entering supported accommodation. This is also supported by public policy. Maintaining involvement in activities which promote well-being or pleasure is known to help offset negative feelings, including those associated with depression (Scogin et al. 2016). However, age-associated health conditions can compromise older adults' abilities to remain engaged in activities outside of the home, which are known to promote well-being, and in consequence, well-being activities become more confined to those within the home-sphere.

Being unable to engage in activities outwith the home can have a negative impact upon an individual's quality of life. It is known that maintaining social networks and participating in activities that promote social interaction are important for well-being (Cornwell and Waite 2009). Lansford et al. (1998) noted that, with increasing age, the number of people in an individual's social network decreases. The oldest old (aged 85+) may have few opportunities to engage in social activities outside of their own homes and, as noted by Kivett et al. (2000), this age group of rural adults have few visits from neighbours and friends. Limiting long-term conditions can compound difficulties in maintaining social connections. Gierveld (1998) observed that older adults can experience loneliness and depression, especially if friends and family live at a distance. Living in a remote and rural area may heighten the risk of older adults with a chronic illness becoming socially isolated. Aspects of contemporary rural life in the Global North such as sparsely populated localities, dispersed settlement structures, accessibility challenges—particularly for non-car drivers and the impacts of long-term migration patterns can often result in older adults having few, if any, close family members living nearby. Long-term residents not knowing their neighbours adds a particular rural dimension to the challenge of maintaining social interaction opportunities for the older population. Combined, these aspects of rural living can increase the likelihood of older rural adults having small social networks and limited opportunities for in-person social interaction. Social interaction with a health or social care provider is, for many older rural adults, the only regular in-person social interaction they have (Farmer et al. 2005). Rural areas offer fewer opportunities for those with a chronic condition to be distracted from the chronic pain that frequently accompanies a long-term limiting illness (Hoffman et al. 2002).

New digital technologies offer a new means of engaging in activities which promote well-being. Many older adults are regular users of digital and Internet-enabled devices and they use these technologies to maintain social connections with friends and family, to interact with their local community and to engage with special interests. There is considerable potential to exploit digital technologies to enhance the well-being of largely and completely housebound older adults. Digital technologies, such as devices that people use to keep in touch with others including mobile (cell) phones, smart phones, computers, Internet-enabled 'smart' televisions and the applications that these technologies can support, are becoming increasingly ubiquitous in everyday life. Email, social networking sites, blogs and voice-over-Internet applications such as SKYPE and Facetime all have a role to play in facilitating social interaction and the promotion of wider well-being at all stages of the life course. They can facilitate social interaction amongst people of all ages and overcome barriers of distance and are thus particularly useful in helping to overcome the social barriers largely housebound older adults can face. Digital devices and/ or applications are also increasingly being used to support older adults to live an active, independent life.

Home healthcare technologies fall into two broad categories: telecare, such as fall alarms and motion sensors which require no direct user input, and telehealth, such as blood pressure, blood glucose and respiratory measures which require active user involvement (Mort and Phillip 2014). Older adults are increasingly likely to use Internet-enabled digital technologies—for health-related or other reasons—in their everyday lives. UK data reports that 50% of older adults aged over 75 use the internet (Ofcom 2017a) and use of digital devices such as tablets and smartphones is increasingly adopted in this age group (Ofcom 2017b). Similarly in Europe, 57% of adults aged 55–74 are regular internet users (Eurostat 2016). Using digital technologies to support aspects of active, independent ageing is now a realistic proposition, especially amongst those aged 65+ and should not be dismissed amongst those aged 85+.

Study Design and Methods

A two-phase design was used in this research project.[2] The first phase collected data that informed the overall concept, design and development of a technology intended to provide opportunities for engaging in the outside world for largely housebound older adults living in rural areas. The second phase was an evaluation of a technology prototype which included a preliminary assessment of how use of the technology could enhance the well-being of its users. The study involved a total of 43 participants. The Phase 1 interviewees ($n = 19$) all lived in North-East Scotland. Sixteen lived in a rural area (as defined under the Scottish Government's urban–rural classification, details available at The Scottish Government 2014), of whom two lived in remote small towns, five in accessible rural areas and nine lived in remote rural areas. The remaining three interviewees lived in the outskirts of Aberdeen, (considered a large urban area) or in the outskirts of areas considered 'other urban'. The Phase 1 focus groups were conducted with 14 older adults living in or near to a large urban area (Dundee). In Phase 2, one-to-one evaluations with were conducted with 10 older adults. Four of whom lived in accessible and remote small towns, three in accessible rural areas and three in remote rural areas of North-East Scotland. The first phase of the study adopted a multiple methods approach involving semi-structured interviews and focus groups. The interviews explored details of the types of sociable and solitary activities interviewees participated in currently and discussed activities the older adults would have liked to be able to participate in, but which their physical health

[2]Project title: Window to the Outside World: Designing a new technology to supplement opportunities for community engagement of older adults in rural NE Scotland.

and mobility constraints prevented them from doing. Findings from the interviews included details of the types of activities that older, largely housebound adults would like to be able to participate in, potentially by using digital technology. The second data collection during Phase 1 involved conducting focus groups with a second group of older adults who were members of the Social Inclusion in the Digital Economy User Group based at the University of Dundee. These focus groups elicited feedback about possible designs for a technology to facilitate engagement with the outside world, which could then be developed into a prototype. Focus group participants discussed interface design and ergonomic features and offered other perspectives on the types of activities older adults might like to be involved into promote engagement with the outside world. In Phase 2, the findings from the Phase 1 exploratory interviews and focus groups were triangulated, leading to an overall concept for the technology, that is, what the overall purpose of the technology should be. A working prototype was then designed, developed and evaluated by ten older adults[3] in one-to-one evaluations, conducted in their homes or another place selected by the participant. The Phase 2 two evaluators either experienced limited mobility themselves, or lived with, knew well or worked closely with others who were housebound. For further details of the study's methods, see Dowds (2016).

Informing the Technology Design

Findings from the Phase 1 interviews provided a rich insight into the social lives and preferences of older, largely housebound adults who lived in rural areas of North-East Scotland. Interviewees were currently and had previously been involved in a wide range of activities. They gave numerous examples of previous involvement in many local activities, with some playing key roles in the establishment of and running of these activities, including, for example, activities for children, sporting events, church activities and opportunities to be out and about in the

[3]Older adults who lived in rural areas North-East Scotland and were either housebound, visited or knew housebound older adults well.

countryside or engaging with nature (such as bird watching) as well as being involved in annual place-specific activities such as local Highland Games. Their level of involvement in activities (e.g. helping to run activities or simply attending them) and the length of time they had been involved in events taking place in their local community varied. The interviewees discussed missing being able to attend and participate in a range of local social activities, such as community groups, attending activities with friends, as well as missing the spontaneous encounters with people known to them and with strangers that routinely occur when you are out and about. They also missed simply being outdoors in their local area, walking, visiting shops and other public places.

Interviewees expressed differing attitudes towards participation in local activities. Some were keen to maintain and create new social ties, yet found that there were few options available for doing so due to reasons such as the death of close friends, being unable to leave home to go to places where one-to-one social encounters arise or difficulties associated with relying on friends who provided them with the transport required to get to activities. Dependence on family members was also discussed as a barrier to being able to attend enjoyable activities. Some interviewees talked about not wanting to ask family members for lifts to activities which were not considered 'necessary', such as medical appointments or grocery shopping, and did not want to feel burdensome to their busy families, a finding that supports previous research (Musselwhite 2017; Ahern and Hine 2012). For others, a lack of motivation to engage in social activities, influenced by numerous factors associated with later life, including the loss of loved ones, self-perceptions of being frail and thus unable to leave the home or not wanting to be seen in public using a mobility aid, created additional barriers to participation. Those interviewees who, for whatever reason did not take part in many out-of-home activities, talked about the importance to them of home-based activities, such as reading, completing crosswords and other puzzles, listening to music, watching television, or cooking and baking. Many interviewees either commented on or took noticeable enjoyment from contemplating the views from their windows, including observing wildlife, passers-by, changing colours of the sky and the views they could see from their home. This has been noted elsewhere, for example, valuing the view from the window at home has been noted to be important for rural older adults (Farmer et al. 2010) and the importance of 'motion' in changing landscapes, watching traffic, etc. for those who are physically 'motionless' (i.e. housebound) has also been found (Musselwhite 2015). Overall, the interview findings provided suggestions about the types of activities that largely housebound older adults would like to have brought into their home via digital technologies.

The focus group discussions provided further examples of the difficulties older adults can face in attempting to remain involved in local social activities. Attitudes expressed about the possibility of using digital technology as an alternative or supplementary means of involvement in activities were discussed favourably. The potential to still be able to experience local activities in some way, despite not being there in-person (i.e. '*moving sideways*' [quote of a focus group participant]) was discussed as being hypothetically more beneficial than having no experience of the activity at all. Notions of 'moving sideways' had also been expressed in interviews with housebound interviewees in Phase 1. For example, a couple who were interviewed, Theresa and William, discussed how meaningful it was for them when members of the community brought 'soup and sweet' to their home, from a community event which they used to play key roles in organising. They felt included despite not being able to attend in-person. Attempts to promote a sense of inclusion, such as being able to watch or take part in local activities from home, could promote well-being and feelings of remaining engaged with the wider world. Furthermore, being able to watch activities from home offers flexibility not possible if events are being attended in-person. For example, footage of local activities could be watched live or at a time more convenient to the older adults. Triangulating the findings from the Phase 1 interviews and focus groups, along with reference to relevant Human–Computer Interaction studies (i.e. those involving designing technologies for older adults), informed the design of a technology prototype. The corresponding technology design and evaluations which comprised Phase 2 of the study will now be discussed.

The overall concept of the technology designed in Phase 2 was to *bring the outside in* (Dowds et al. 2015), through enabling older adults with a new avenue to be able to maintain involvement in local activities.

Key elements to incorporate into the technology design included: being able to watch a wide range of local activities remotely, and being able to watch live and recorded footage. The prototype technology thus aimed to accommodate the desire of many older adults to remain engaged with their local community and the wider world beyond their home through being able to watch footage of events and activities that they would have once have participated in in-person, in the same way as they would tune into a television programme. The prototype was then designed to respond to these identified requirements, facilitating a means of engaging in a range of activities from the home. An overarching aim was that the prototype would be simple-to-use technology. It would be capable of streaming live footage of local activities providing a 'portal', which could enable older, housebound adults to watch activities in which familiar places or people were seen. Guidelines routinely adopted in Human–Computer Interaction studies were applied, specifically recommendations about the design and development of technology for use by older adults (such as using a large font size, limiting the amount of text on the screen, including simple to navigate functions, e.g. Doyle et al. 2010, as well as using a digital device suitable for use by people with, for example, limited dexterity in their hands as described by Chen and Chan 2013). A tablet computer was selected as the most appropriate type of digital device to support the technology concepts, and software for use on a tablet was programmed by a Computing Science colleague of the research team.

The prototype technology was called *Window to the Outside World*. The term 'window' was adopted to echo the enjoyment many interviewees had described gaining from looking out from windows in their home. The word 'window' also encompassed the sentiment of being able to view *local* activities, which although may not be visible from views from their homes, were known to be taking place close by and in a familiar setting, such as the village hall or town square. The types of activities which the Phase 1 interviewees mentioned that they missed were grouped into four categories: 'Services', 'Nature', 'Events' and 'Street', and each category was represented as a 'window' image on the prototype (see Fig. 5.1). Example footage was recorded in and around Aberdeen, adhering to the British Broadcasting Corporation guidelines

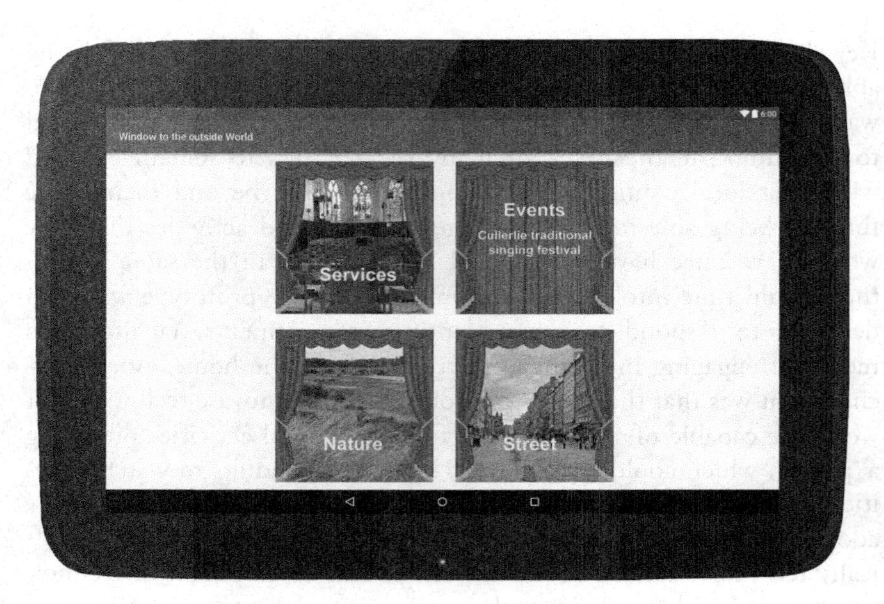

Fig. 5.1 Screenshot of the main menu of the Window to the Outside World prototype

on filming in the public arena, whereby 'individuals may be captured, yet will not purposefully be focused on unless they provide consent' (British Broadcasting Corporation 2015). The footage included: a church service, a street scene, a beach film, a garden film, a choir concert and a singing festival.

The aim of the technology was to provide live footage of local events to the Phase 2 evaluators. However, due to constraints related to internet availability, bandwidth requirement and challenges associated with establishing live broadcasts to the homes of all Phase 2 evaluators, the prototype was uploaded with pre-recorded footage. The uploaded footage was not necessarily local or familiar to the older adult evaluators in the second phase; however, it provided an example of how the technology could work. Having the *possibility* of being able to watch events which take place in the local community of the user was investigated in the evaluations. Details of the Phase 2 evaluations of the prototype now follow.

Evaluating the Technology

The Phase 2 evaluations of the prototype involved firstly a brief exploration of the evaluators' abilities to get out and about and their current and previous involvement in the local community and, secondly, demonstrating how the prototype would be used and then obtaining evaluators' feedback on the overall concept of the prototype technology *Window to the Outside World*. The second part of the evaluations involved showing interviewees the technology and uploaded footage. Therefore, feedback was elicited on the technology itself and the illustrative footage, as well as other aspects of the user experience, including how easy the prototype technology was to use, and whether it was perceived to be useful. An overview of the evaluation phase findings follows. Three particular themes arose from the discussions with the older adult evaluators, namely: the perceived usefulness of the prototype; opinions relating to being able to watch either live or recorded footage; and the availability of local or non-local footage.

The concept of being able to keep up-to-date with local activities using the technology was well-received by those who participated in the evaluation. In fact, two participants were already keeping up-to-date with local activities through technology. One interviewee watched DVDs of recent events such as the local pantomime or Christmas carol concerts filmed by a local resident and another liked to see photographs that her husband often took of local changes such as new buildings or shops and events in the area in which she lived.

The merits of being able to watch live or recorded footage were discussed. The technology was originally intended to stream live footage only. Live footage was discussed with reference to inducing a sense of 'togetherness' which may be less frequently experienced when housebound, through watching live events or simply knowing that others would be watching the same footage in their homes at the same time (Sokoler and Svensson 2007). An opportunity to view live footage could allow older people to uphold a sense of engagement with past routines: perhaps doing things they used to like to do, such as singing along with church service hymns from home whilst watching the service live.

Additionally, the flexibility to watch footage as and when desired was discussed. Unlike live streaming, which would be available for the duration of the activity taking place, recorded footage can accommodate the choices and preferred watching times of the viewers. These preferences may be influenced by health-associated factors, such as tiredness or carers visiting times, or other factors such as hosting visitors. Furthermore, evaluators of the technology noted that the times when people might use the device would depend on factors such as if an older adult felt bored, if they felt unwell or if there was nothing they wanted to watch on the television. Overall, a clear preference was expressed for the technology to incorporate opportunities to watch both live and recorded footage.

Respondents discussed the importance to them of being able to watch local activities. The many benefits included, for example, the ability to follow up on local news stories and to keep up-to-date with happenings in the local community. Others wished to continue to pursue interests rooted in the place where they lived and which might have influenced their decision to reside in that particular place. For example, opportunities to go bird watching, to walk at coastal locations or to see developments in woodland bought by the local community were cited as being important to some people. Thus, having the ability to view these types of scenes, particularly if the viewer was no longer able to physically visit them was considered to be beneficial, as opposed to having either very limited, or no access whatsoever. The ability to watch local activities and local places also increased the likelihood of recognising familiar faces and places and re-living everyday activities they were no longer able to perform, as Donald echoed:

> Coz everybody likes to wander up and down the shops…Just to see would be a great bonus. But, if it was live on a Saturday morning, you could see who's out and about…I could imagine people sitting all morning just sitting watching who's going up and down the high street! (Donald, prototype evaluator, aged 75–84)

Some Phase 2 evaluators who considered that there was a dearth of local activities taking place in their community displayed a preference to be able to watch activities which were not necessarily local, yet perhaps used to take place locally (such as traditional Scottish music or shows/drama productions) or those that they particularly enjoyed yet had limited access to.

Phase 2 evaluators discussed preferences for footage that they could directly relate to, such as sports or hobbies they used to partake in, scenic views or walks they used to enjoy, or community activities in which they used to play a role. This aligned with the findings in Phase 1 where interviewees commented on their enjoyment of re-living experiences from their past. For example, Rose mentioned reading books set in places where she used to live, as a means of taking an imaginative tour around sights and places she was familiar with, even despite not necessarily liking the author. Edna enjoyed watching nature programmes which showed particular flowers (including rare varieties) or wildlife that she had previously witnessed first-hand. Thus, the activity of bringing certain experiences from the past into the present tense, through reading or watching the television, was enjoyed and the potential to use the technology as a supplementary means of being able to tune into familiar or enjoyable activities was well-received. Furthermore, although respondents enjoyed re-living experiences from their past through familiar or meaningful stimulus, they also were appreciative of the fact that the technology could provide a snapshot of activities taking place currently. The technology could provide an unedited replica of being in the outside world, including the current seasonality witnessed through the natural flora and fauna at that time of year, as well as allowing the viewer to observe natural social interactions such as free-flowing conversations. To summarise, the footage showing content that was familiar to viewers (and therefore provided a stimulus to revive personal experiences), combined with the prototype offering an ability to view current sights and places where they were physically unable to go in-person, was well-received by the evaluators.

Discussion

The findings from both phases of the research raise several important points for discussion, primarily the importance of facilitating the maintenance of meaningful connections for older, housebound adults living in rural areas. This section will consider this alongside the potential for technology such as the prototype *Window to the Outside World* to stimulate imaginative mobility. Through reviving and maintaining meaningful connections and attachments, there is considerable potential for the technology to initiate and support behaviour that promotes well-being, such as identity building and access to enjoyable footage, which may enhance one's sense of pleasure. There is also scope for further benefits to be gained through the independence associated with using such a technology, rather than relying on others, for keeping connected with the outside world.

Having and maintaining a sense of engagement with the outside world, particularly through activities that respondents held a personal connection to, were discussed favourably by Phase 1 and 2 participants. The importance of personal relationships or bonds held between people and aspects of the world around them has been the focus of enquiry in a number of academic disciplines. Reflections on *place attachment* offered by academics drawing on the findings of UK-based research are such an example. Place attachment studies address the metaphorical 'glue' that connects people with places and have focused on rural-based older adults in particular (e.g. Burholt 2006; Hennessy et al. 2014). Over time, daily and historical experiences of living in a specific area, such as the people one meets or the buildings or open spaces one passes or inhabits, can endow an individual with a dynamic map of connections, associations and emotions that they hold with a particular place. These feelings or connections associated with places can be heightened, strengthened or 'nourished', through daily encounters with others (Hennessy et al. 2014: 15), for reasons including the reinforcement of one's sense of membership in the community and inclusion in an identifiable social group (Burholt 2006). Maintaining this relationship has been identified as playing an integral role in the well-being of older adults (e.g. Hennessy et al. 2014). For largely housebound older adults, the ability to participate in the type of encounters that are associated with 'nourishing' these attachments may be particularly limited, such as through getting out and about to meet people. We posit that for many largely housebound rural older adults, the importance of possessing forms of connections with the outside world is salient; however, the opportunities for these to be stimulated and potentially strengthened are limited.

Other than social encounters, we suggest that another means of stimulating and potentially strengthening or nourishing these important connections and attachments is through imaginative mobility, which use of the technology could stimulate. As physical mobility reduces for older adults, it is proposed that alternative 'ideational' forms of mobility become more salient, including virtual and imaginative mobility (Parkhurst et al. 2014: 22). The concepts of virtual and imaginative travel were introduced by Urry (2004), which broadened the traditional mobilities paradigm, recognising forms of 'metaphorical' travel (Frello 2008: 27) in addition to physical or corporeal travel. These concepts have since become expanded. For example, the term 'virtual mobility' refers to the use of information and communication technology to experience virtually 'visiting' a location out-of-home and has previously been discussed with regard to older adults (Parkhurst et al. 2014). Similarly, accounting for the psychological aspects of imaginative travel, this concept has become broadened and is now more commonly referred to as 'imaginative mobilities'. Ziegler and Schwanen (2011: 775) interpret this as 'cognitive processes (memory and imagination) that recollect or construct events in other times (past or future) and other places'. Parkhurst et al. (2014: 11) define imaginative mobilities as:

> …the ways in which people extend their sense of connectedness to, and meaningful engagement with, life activities that were previously addressed by corporeal mobility. This [involves] […] the achievement of a different but allied experience through other means (such as reminiscing about long-standing friends and stories with visitors and through photographs).

We posit that the technology introduced in this chapter will provide an opportunity for virtual mobility, which could then stimulate imaginative mobility, by enhancing one's ability to extend their sense of connectedness or meaningful engagement with activities or places, which hold an important personal meaning and that are difficult to physically visit. Being able to watch current footage of familiar activities or places through technology such as *Window to the Outside World* could lead to the 'nourishing' or enrichment of memories or experiences associated with the past, giving the older adult increased opportunities for an 'allied' experience, which brings together the present, with their past in a new way. Thus, we propose that using the technology could expand one's access to meaningful activities or places (both local and non-local), which could then stimulate imaginative mobility and bring potential corresponding benefits to well-being. These include reinforcing a positive sense of personal identity though promoting continuity, a general sense of pleasure and therapeutic benefits, which will now be discussed.

Sense of Personal Identity

The findings suggest that watching footage of local activities which are personally meaningful to older adults (due to reasons including previous long-term participation, involvement or familiarity) could offer a sense of continuity to the identities of older adults. The Phase 1 interviewees associated themselves with identities that were often related to their (previous) occupation or interests that they took pride in. For example, Dorothy discussed with pride how she was known locally for her musical talents. Schwanen and Ziegler (2011) observed that later life can bring challenges (including being less able to get out and about if one has to give up driving) in maintaining one's sense of identity, which can have negative implications for the well-being of older adults. Therefore, creating more resources for enabling older adults to remain connected with an aspect of their positively perceived self-identity could bring benefits. Ageing in place, be it in one's own home and/or ageing in a familiar community, enables one's sense of identity to be reinforced by familiar belongings and living in a familiar setting (Stones and Gullifer 2016). Maintaining a sense of continuity, specifically internal continuity, which relates to the structures of one's sense of self and identity (Atchley 1989), has been found to be beneficial for older adults as an adaptive way of coping with change. The familiarity associated with watching favourite television programmes has been found to help enhance a sense of continuity amongst older adults at times of transition or change (Van Der Goot et al. 2012). In essence, Dorothy may gain particular enjoyment from watching a live-streamed music concert held at the local pub that she used to perform in herself. Watching this may evoke positive memories of a previous time of her life and consequently reinforce her positively perceived sense of identity associated with being a local musician. Our study suggests that having a means of watching local, familiar or meaningful activities and leisure pursuits which had previously played an important role in the personal identity of older adults, rather than having no means at all, could be beneficial for well-being when physical health causes various transitions that inhibit one's ability to remain involved in these types of local activities.

Pleasure

The personal interests and preferences of older adults are important indicators of the types of footage that would be meaningful to watch on technology such as *Window to the Outside World*. Being able to re-live familiar routines through the technology, such as walking down the high street at the weekend and being able to potentially recognise familiar faces, may provide a new enjoyable pursuit which can take place from one's home. However, it should not be assumed that older peoples' interests are confined to activities and events that are available in the local community they live in. They are likely to be interested in activities taking place in other localities, including those they lived in in the past. For example, Phase 2 evaluator Patricia was keen to watch live performances of traditional Scottish music event which no longer took place in her own locality. Her preference infers that enjoyment is likely to be gained from being able to be engaged, through the technology, with activities that older adults hold a personal relationship with, such as the types of activities they used to enjoy attending or participating in. We propose that the nature of the 'allied' experience of imaginative mobility would be richer, perhaps with a heightened vividness of associated sensory memories or the likelihood of valuable and meaningful memories being stimulated as opposed to memories with little accompanying affect, and thus more enjoyable, when watching an activity that the older adult holds some sort of personal association with.

Recent research highlights the importance of involvement in *pleasant events* for rural older adults (Scogin et al. 2016). Involvement in activities which bring a sense of enjoyment or meaning is expected to offset negative emotions such as depression, which is more common amongst rural older adults (ibid.) and increases in likelihood with age, particularly amongst those with a physical illness (Barnett et al. 2012). As previously mentioned, intertwining barriers contribute to challenges with engaging in these types of activities regardless of whether they take place locally or further afield. These challenges include limited access to transport for leisure activities and relying on others for lifts. Therefore, technology such as *Window to the Outside World* may offer an important new and/or only means of accessing activities associated with personal meaning or enjoyment for those with few friends or family living nearby to provide transport. One's ability to watch footage of activities (including those which no longer take place locally) or which represent familiarity and are associated with enjoyment or pleasure, may be increased through the use of such technology.

Therapeutic Benefits

Aligning with the preference to witness natural and unedited footage was the inference of benefits from watching footage of these types of scenes. For example, some Phase 2 evaluators mentioned the therapeutic effect of watching some featured scenes, specifically those in the *Window to the Outside World* 'Nature' footage, such as the movement of water. It is worth noting that 'seeing beautiful scenery' or 'listening to the sounds of nature' are included as indicators as 'Pleasant Events' in the study mentioned in Rider et al. (2016). The benefits of spending time in outdoor, 'green' and 'blue' spaces, for older adults (including those who live in rural areas) have been reported (Finlay et al. 2015; Philips 2014; Butler and Cohen 2010), as have the barriers to outdoor access, which are more prominent for older people than for younger age groups (Colley et al. 2016). The fact that the footage of the 'nature' window sparked comments relating to a sense of comfort or relaxation infers that watching natural, outdoor scenes remotely may be beneficial for largely housebound older adults whose physical access to the natural world is limited. Previous research indicates that technology cannot transcend the benefits gained from watching outdoor scenes from a window (Kahn et al. 2008). However, evidence suggests that watching live footage of nature can bring psychological benefits yet this requires further study (Friedman et al. 2008). The interviewees were very fond of the views seen from the windows of their homes. It may be worthwhile expanding the choice of available footage to include images of places they are no longer able to physically travel to. For example, images or videos of places that cannot be easily accessed from home, such as local or familiar lakes or forest walks, could be beneficial for viewers due to the potential stimulation of personal reflections and memories associated with the local or familiar and thus, meaningful footage.

Independence

The ability to view local activities from home through use of technology such as *Window to the Outside World* may overcome barriers that limit physically taking part in activities, such as relying on others to be taken to places, events and activities. As mentioned, these challenges include perceiving oneself as a burden (Ahern and Hine 2012) or difficulties with relying on friends of a similar age with similar health problems (see Dowds 2016). Use of the technology could provide flexibility to watch and engage with activities which the normal challenges associated with physically attending activities (and thus relying on other people) may otherwise prevent. The simple interface used in *Window to the Outside World* was designed to enable older adults, including those who are not familiar with tablets, to use the technology unassisted. Therefore, aside from assistance from others to learn how to use the technology initially (which many older adults, being familiar with tablets, would not require), engaging with local activities through *Window of the Outside World* would provide a means of keeping up-to-date with the local community, without relying on others. Having first-hand access to local information through watching live-streamed footage has the potential to increase the viewers' sense of independence and to empower them with local knowledge that they would otherwise be dependent on others to learn (such as friends or health care professionals, see Philip et al. 2015). This may engender older adults with a sense of control which they previously lacked and represents a means of circumventing increased dependence on others common to later life, a dependence which typically detracts from personal feelings of control, potentially impacting on their well-being (Stones and Gullifer 2016; Ward 2012).

Conclusion

The premise of the *Window to the Outside World* project was to '*bring the outside in*' to largely housebound rural older adults. In doing so, viewers would be presented with a virtual means of seeing familiar places and watching local activities that they would be expected to hold some sort of connection with (due to, e.g., living in the local area where the footage was recorded). Such remote access to familiar and meaningful footage carries the potential to expand one's personal resources for stimulating imaginative mobility, reinforcing a sense of connection with spaces that hold emotional or physical meanings for them outside the confines of their home. Where social encounters with others can 'nourish' existing place attachment ties, we propose that being able to regularly watch footage of familiar sights and scenes could also reinforce and enrich these important connections. Many of the interviewees and evaluators had previously held long-standing roles in their communities either in the area where they currently lived or where they had previously lived. The prospect of continuing a sense of engagement in the communities they were once vibrant members of may be particularly beneficial for older adults. Using the technology as a supplementary interest to the television may be beneficial for purposes of distraction for those with physical health problems in rural areas, which has been identified as beneficial, for example, for those who suffer from chronic pain (Hoffman et al. 2002). Furthermore, the continuity benefits and upholding meaningful connections may be particularly useful at times of transition in later life, including the sudden or gradual disengagement from activities outwith the home which commonly accompanies limited mobility. Our research findings infer that digital technology that facilitates visual and audio access to familiar or meaningful places could support well-being amongst older adults who are largely housebound and living in rural areas.

The delivery of clinical services through the use of telemedicine could be expanded to incorporate aspects of *Window to the Outside World* as a means of bolstering overall well-being for older adults who are residing at home, yet whose physical mobility is limited. Use of such technology is not designed nor intended to replace any opportunities to physically take part in local activities and socialise. Instead, it could offer a supplementary means of reviving connections that may lie dormant due to physical restrictions. It could instigate pleasure as well as other potential benefits, such as identity continuity and supplementary means of gaining benefits from nature which largely housebound older adults may otherwise have limited access to. This type of technology could broaden access to the natural sights that interviewees enjoyed currently observing from their homes. It presents an opportunity to engage with current local activities, whilst enriching meaningful connections from the past through imaginative mobility. Increasing numbers of older adults are adopting the use of tablets, thus the expectations of what they should be able to do (along with the ever-expanding range of what technology can allow users to do) will undoubtedly also increase. Although the use of social media can provide real-time access to the outside world (such as the new 'Facebook live' feature), a simple to use interface and a variety of local footage and footage of the types of things that hold significance for the viewers are important for improving place attachment.

Collaborations with local stakeholders could support successful rollout of the technology and provide an opportunity for longitudinal evaluation amongst a larger group of users. Further developments of the prototype could be considered, such as a facility to tune into live local activities. Some evaluators did not foresee wanting to use the technology to participate remotely in live activities (such as performing a church reading from home), yet older adults whose involvement in local activities ends abruptly rather than gradually due to a sudden change in their health status, may appreciate having an option to remain engaged remotely and having a means of participating as opposed to none. Tuning into live activities could also enable real-time interaction between users of the technology and other viewers, or those physically attending the local activity. In follow-up research, it would be worth investigating whether use of such technology would in fact lead to increased social interactions for the viewer, and whether these virtual interactions would enhance health benefits associated with maintaining one's social network (e.g. Ashida and Heaney 2008; Cornwell and Waite 2009). Careful attention should be paid to whether the use of such technology may subsequently decrease the number of in-person social interactions experienced. This may become an unintended consequence of use of such technology, if virtual interactions were substituted for in-person interactions, which are particularly important for older adults living in rural areas, with chronic illnesses (e.g. Corbett and Williams 2014). Similarly, use of such technology would not be intended to substitute in-person visits to local areas or events, if this remained possible for the older adult viewers. Future improvements of this type of technology should cater for providing as many options as possible, such as live and recorded, as well as local and non-local footage, whilst focusing first and foremost on ways to facilitate engagement with (even if it is purely watching) meaningful activities.

Acknowledgements We would like to thank all of the participants who took part in the research project, Dr. John Paul Vargheese who programmed the working prototype, Alex Tang and Dr. Annie McKee for their thorough, helpful comments on an earlier draft of this chapter. This project was supported by the award made by RCUK Digital Economy Theme to the dot.rural Digital Economy hub, award reference: EP/G066051/1. Further acknowledgment is to the SiDE User Panel at the University of Dundee, awarded by RCUK Digital Economy Research Hub EP/G066019/1—SIDE: Social Inclusion through the Digital Economy.

References

AgeUK. 2015. Briefing: The Health and Care of Older People in England 2015. http://www.cpa.org.uk/cpa/docs/AgeUK-Briefing-TheHealthandCareofOlderPeopleinEngland-2015.pdf. Accessed 16 March 2017.

Ahern, A., and J. Hine. 2012. Rural Transport—Valuing the Mobility of Older People. *Research in Transportation Economics* 24: 27–34. http://dx.doi.org/10.1016/j.retrec.2011.12.004. [Online].

Ashida, S., and C.A. Heaney. 2008. Differential Associations of Social Support and Social Connectedness with Structural Features of Social Networks and the Health Status of Older Adults. *Journal of Aging and Health* 20: 872. https://doi.org/10.1177/0898264308324626. [Online].

Atchley, R. 1989. A Continuity Theory of Normal Ageing. *The Gerontological Society of America* 29 (2): 183–190. https://doi.org/10.1093/geront/29.2.183. [Online].

Barnett, K., S.W. Mercer, M. Norbury, G. Watt, S. Wyke, and B. Guthrie. 2012. Epidemiology of Multimorbidity and Implications for Health Care, Research, and Medical Education: A Cross-Sectional Study. *Lancet* 7 (380): 37–43. https://doi.org/10.1016/s0140-6736(12)60240-2. [Online].

Blake, S. 2009. Subnational Patterns of Population Ageing. *Population Trends* 136 (Summer): 43–63.

British Broadcasting Corporation. 2015. Editorial Guidelines: Section 7 Privacy. http://www.bbc.co.uk/editorialguidelines/page/guidelinesprivacy-privacy-consent/. Accessed 11 March 2015. [Online].

Burholt, V. 2006. 'Adref': Theoretical Contexts of Attachment to Place for Mature and Older People in Rural North Wales. *Environment and Planning A* 38: 1095–1114. https://doi.org/10.1068/a3767. [Online].

Butler, S.S., and A.L. Cohen. 2010. The Importance of Nature in the Wellbeing of Rural Elders. *Nature and Culture* 5 (2): 150–174. https://doi.org/10.3167/nc.2010.050203. [Online].

Chen, K., and A.H. Chan. 2013. Use or Non-use of Gerontechnology—A Qualitative Study. *International Journal of Environmental Research and Public Health* 10 (10): 4645–4666. https://doi.org/10.3390/ijerph10104645. [Online].

Colley, K., M. Currie, J. Hopkins, and P. Melo. 2016. Access to Outdoor Recreation by Older People in Scotland. http://www.gov.scot/Publications/2016/06/8917/downloads#res500263. Accessed 16 February 2017. [Online].

Corbett, S., and F. Williams. 2014. Striking a Professional Balance: Interactions Between Nurses and Their Older Rural Patients. *British Journal of Community Nursing* 19 (4): 162–167. http://dx.doi.org/10.12968/bjcn.2014.19.4.162. [Online].

Cornwell, E.Y., and L.J. Waite. 2009. Social Disconnectedness, Perceived Isolation, and Health Among Older Adults. *Journal of Health and Social Behaviour* 50 (1): 31–48. https://doi.org/10.1177/002214650905000103. [Online].

Damant, J., and M. Knapp. 2015. What are the Likely Changes in Society and Technology Which Will Impact Upon the Ability of Older Adults to Maintain Social (Extra-Familial) Networks of Support Now, in 2025 and in 2040? Future of an Ageing Population: Evidence Review. Government Office for Science. https://www.gov.uk/government/uploads/system/uploads/attachment_data/file/463263/gs-15-6-technology-and-support-networks.pdf. Accessed 26 January 2015.

Dowds, G.L. 2016. Window to the Outside World: Designing a New Technology to Supplement Opportunities for Community Engagement of Older Adults in Rural NE Scotland. http://digitool.abdn.ac.uk:80/webclient/DeliveryManager?application=DIGITOOL-3&owner=resourcediscovery&custom_att_2=simple_viewer&pid=231023. Available from the University of Aberdeen Library and Historic Collections Digital Resources. Accessed 15 March 2017. [Online].

Dowds, G., L.J. Philip and M. Currie. 2015. Bringing the Outside In: Technology for Increasing Engagement with the Outside World among Rural Housebound Older Adults. The XXVI European Society for Rural Sociology Congress, August 2015, Aberden, Scotland, at http://www.esrs2015.eu/sites/www.esrs2015.eu/files/ESRS%202015%20on-line%20proceedings.pdf. [Online].

Doyle, J., Z. Skrba, R. McDonnell, and B. Arent. 2010. Designing a Touch Screen Communication Device to Support Social Interaction Amongst Older adults. In *Proceedings of the 24th BCS International Conference on Human-Computer Interaction*, 177–185, September 6–10, Dundee, United Kingdom.

Eurostat. 2016. Internet Access and Use Statistics—Households and Individuals. http://ec.europa.eu/eurostat/statistics-explained/index.php/Internet_access_and_use_statistics_-_households_and_individuals#Further_Eurostat_information. Page updated 30 January 2017. Accessed 22 September 2017. [Online].

Farmer, J., L. Philip, G. King, J. Farrington, and M. Macleod. 2010. Territorial Tensions: Misaligned Management and Community Perspectives on Health Services for Older People in Remote Rural Areas. *Health & Place* 16 (2): 275–283. http://dx.doi.org/10.1016/j.healthplace.2009.10.010. [Online].

Farmer, J., C. West, B. Whyte and M. Maclean. 2005. Primary Health-Care Teams as Adaptive Organisations: Exploring and Explaining Work Variation Using Case Studies in Rural and Urban Scotland. *Health Services Management Research* 18 (3): 151–164. https://doi.org/10.1258/0951484054572501. [Online].

Finlay, J., T. Franke, H. Mckay, and J. Sims-Gould. 2015. Therapeutic Landscapes and Wellbeing in Later Life: Impacts of Blue and Green Spaces for Older Adults. *Health & Place* 34: 97–106. http://dx.doi.org/10.1016/j.healthplace.2015.05.001. [Online].

Frello, B.F. 2008. Towards a Discursive Analytics of Movement: On the Making and Unmaking of Movement as an Object of Knowledge. *Mobilities* 3 (1): 25–50. https://doi.org/10.1080/17450100701797299. [Online].

Friedman, B., N.G. Freier, and P.H. Kahn, Jr. 2008. Office Window of the Future?—Two Case Studies of an Augmented Window. In Extended Abstracts of SIGCHI Conference on Human Factors in Computing Systems Vienna, Austria, 1559. https://doi.org/10.1145/985921.986135.

Gierveld, J. 1998. A Review of Loneliness: Concepts and Definitions, Determinants and Consequences. *Reviews in Clinical Gerontology* 8 (1): 73–80. https://doi.org/10.1017/S0959259898008090.

Hennessy, C.H., Y. Staelens, G. Lankshear, A. Phippen, A. Silk, and D. Zahra. 2014. Rural Connectivity and Older People's Leisure Participation. In *Countryside Connections: Older People, Community and Place in Rural Britain*, ed. C.H. Hennessy, R. Means, and V. Burholt. Policy Press Scholarship. [Online].

Hoffman, P.K., B.P. Meier, and J.R. Council. 2002. A Comparison of Chronic Pain Between an Urban and Rural Population. *Journal of Community Health* 19 (4): 213–224. https://doi.org/10.1207/s15327655jchn1904_02. [Online].

International Council on Active Ageing. 2001–2016. Overview: The ICAA Model. http://www.icaa.cc/activeagingandwellness.htm. Accessed 09 February 2016.

Kahn Jr., P.H., B. Friedman, B. Gill, J. Hagman, R.L. Severson, N.G. Freier, E.N. Feldman, S. Carrère, and A. Stolyar. 2008. A Plasma Display Window?—The Shifting Baseline Problem in a Technologically Mediated Natural World. *Journal of Environmental Psychology* 28 (2): 192–199. http://dx.doi.org/10.1016/j.jenvp.2007.10.008. [Online].

Kivett, V.R., M.L. Stevenson, and C.H. Zwane. 2000. Very-Old Rural Adults: Functional Status and Social Support. *Journal of Applied Gerontology* 19 (1): 58–77. https://doi.org/10.1177/073346480001900104. [Online].

Lansford, J., A. Sherman, and T. Antonucci. 1998. Satisfaction with Social Networks: An Examination of Socio-Emotional Selectivity Theory Across Cohorts. *Psychology and Ageing* 13 (4): 544–552. https://doi.org/10.1037/0882-7974.13.4.544.

Lozano, R., M. Naghavi, K. Foreman, S. Lim, K. Shibuya, V. Aboyans, et al. 2012. Global and Regional Mortality from 235 Causes of Death for 20 Age Groups in 1990 and 2010: A Systematic Analysis for the Global Burden of Disease Study 2010. *Lancet* 380 (9859): 2095–2128.

Mort, A.J. and L.J. Philip. 2014. Social Isolation and the Perceived Importance of In-person Care Amongst Rural Older Adults with Chronic Pain: A Review and Emerging Research Agenda. *Journal of Pain Management* 7 (1): 13–21. http://search.proquest.com/openview/d5e9caf9a2b4c1e-7b8140ec58be3db13/1?pq-origsite=gscholar. [Online].

Musselwhite, C. 2015. The Importance of Motion for the Motionless. The Significance of a Room with a View for Older People with Limited Mobility. International Association of Gerontology and Geriatrics 8th Congress. http://www.drcharliemuss.com/uploads/1/2/8/0/12809985/musselwhite_motion_for_the_motionless_iagg.pdf. Accessed 21 August 2016. [Online].

Musselwhite, C. 2017. Exploring the Importance of Discretionary Mobility in Later Life. *Working with Older People* 21 (1): 49–58. https://doi.org/10.1108/wwop-12-2016-0038. [Online].

National Institute On Aging. 2011. Global Health and Aging. http://www.nia.nih.gov/sites/default/files/global_health_and_aging.pdf. Accessed 17 March 2015.

Nordbakke, S., and T. Schwanen. 2014. Well-being and Mobility: A Theoretical Framework and Literature Review Focusing on Older People. *Mobilities* 9 (1): 104–129. https://doi.org/10.1080/17450101.2013.784542. [Online].

Ofcom. 2017a. Internet Use and Attitudes: 2017 Metrics Bulletin. https://www.ofcom.org.uk/__data/assets/pdf_file/0018/105507/internet-use-attitudes-bulletin-2017.pdf. Accessed 22 September 2017. [Online].

Ofcom. 2017b. Adults' Media Use and Attitudes. Rise of the Social Seniors Revealed. https://www.ofcom.org.uk/__data/assets/pdf_file/0020/102755/adults-media-use-attitudes-2017.pdf. Accessed 22 September 2017. [Online].

Parkhurst, G., K. Galvin, C. Musselwhite, J. Philips, I. Shergold, and L. Todres. 2014. Beyond Transport: Understanding the Role of Mobilities in Connecting Rural Elders in Civic Society. In *Countryside Connections: Older People, Community and Place in Rural Britain*, ed. C.H. Hennessy, R. Means, and V. Burholt. Policy Press Scholarship Online.

Philip, L., D.L. Brown, and A. Stockdale. 2012. Demographic Ageing in Rural Areas: Insights from the UK and US. In *Rural Transformations and Rural Policies in the US and UK*, ed. M. Shucksmith, D. Brown, S. Shortall, J. Vergunst and M. Warner, 58–78. London: Routledge.

Philip, L.J., E.A. Roberts, M. Currie, and A.J. Mort. 2015. Technology for Older Adults: Maximising Personal and Social Interaction: Exploring Opportunities for eHealth to Support the Older Rural Population with Chronic Pain. *Scottish Geographical Journal*. https://doi.org/10.1080/14702541.2014.978806. [Online].

Philips, M. 2014. Baroque Rurality in an English Village. *Journal of Rural Studies* 33: 56–70.

Rider, K.L., L.W. Thompson, and D. Gallagher-Thompson. 2016. California Older Persons Pleasant Events Scale: A Tool to Help Older Adults Increase Positive Experiences. *Clinical Gerontologist* 39 (1): 64–83.

Schwanen, T., and F. Ziegler. 2011. Wellbeing, Independence and Mobility: An Introduction. *Ageing and Society* 31 (5): 719–733. http://dx.doi.org/10.1017/S0144686X10000498. [Online].

Scogin, F., M. Morthland, E.A. Dinapoli, M. Larocca, and W. Chaplin. 2016. Pleasant Events, Hopelessness, and Quality of Life in Rural Older Adults. *The Journal of Rural Health* 32 (1): 102–109. https://doi.org/10.1111/jrh.12130. [Online].

Sokoler, T. and M.S. Svensson. 2007. Embracing Ambiguity in the Design of Non-stigmatizing Digital Technology for Social Interaction Among Senior Citizens. *Behaviour and Information Technology* 26 (4): 297–307. http://dx.doi.org/10.1080/01449290601173549. [Online].

Spinney, J.E.L., D.M. Scott, and K.B. Newbold. 2009. Transport Mobility Benefits and Quality of Life: A Time-use Perspective of Elderly Canadians. *Transport Policy* 16 (1): 1–11.

Stones, D. and J. Gullifer. 2016. 'At Home It's Just So Much Easier to be Yourself': Older Adults' Perceptions of Ageing in Place. *Ageing and Society* 36 (3): 446–481. https://doi.org/10.1017/s0144686x14001214. [Online].

The Scottish Government. 2010. Long Term Conditions Collaborative: Improving Care Pathways. http://www.gov.scot/Resource/Doc/309257/0097421.pdf. Accessed 16 February 2017. [Online].

The Scottish Government. 2013. Rural Scotland Key Facts 2012. http://www.scotland.gov.uk/Publications/2012/09/7993. Accessed 03 March 2015. [Online].

The Scottish Government. 2014. Urban/Rural Classification 2013/2014. http://www.gov.scot/Resource/0046/00464780.pdf. Accessed 27 May 2015. [Online].

United Nations. 2001. World Population Ageing: 1950–2050, Economic and Social Affairs Population Division, ST/ESA/SER.A/207, United Nations, New York. http://www.un.org/esa/population/publications/worldageing19502050/pdf/preface_web.pdfOpenURL, University of Aberdeen. [Online].

United States Census Bureau. 2016. An Aging World: 2015. https://www.census.gov/library/publications/2016/demo/P95-16-1.html. Accessed 22 September 2017. [Online].

Urry, J. 2004. Connections. *Environment and Planning D: Society and Space* 22: 27–37. https://doi.org/10.1068/d322t. [Online].

Van Der Goot, M., W.J. Beentjes, and M. Van Selm. 2012. Meanings of Television in Older Adults; Lives: An Analysis of Change and Continuity in Television Viewing. *Ageing and Society* 32 (1): 147–168. https://doi.org/10.1017/s0144686x1100016x. [Online].

Ward, M.W. 2012. Sense of Control and Sociodemographic Differences in Self-reported Health in Older Adults. *Quality of Life Research* 21 (9): 1509–1518.

World Health Organization. 2002. Active Ageing: A Policy Framework. http://whqlibdoc.who.int/hq/2002/WHO_NMH_NPH_02.8.pdf?ua=1. Accessed 12 March 2017. [Online].

World Health Organization. 2011. Global Health and Ageing. Geneva: WHO. http://www.who.int/ageing/publications/global_health.pdf. Accessed 16 February 2017. [Online].

Ziegler, F. and T. Schwanen. 2011. I Like to Go Out to be Energised by Different People: An Exploratory Analysis of Mobility and Well-being in Later Life. *Ageing & Society* 31: 758–781. https://doi.org/10.1017/S0144686X10000498. [Online].

Part III

Urban

6

Cycling for Transport Among Older Adults: Health Benefits, Prevalence, Determinants, Injuries and the Potential of E-bikes

Jelle Van Cauwenberg, Bas de Geus and Benedicte Deforche

Introduction

Increasing the levels of active transport (walking and cycling) is an important objective of several (inter)national organizations and policies to combat climate change, air pollution and traffic congestion (European Parliament 2010; Biton et al. 2014; Austroads 2010; Thiry 2011). From an individual health perspective, walking and cycling for transport offers an opportunity to integrate physical activity into the daily lives of older adults, who are the least physically active age group

J. Van Cauwenberg (✉) · B. Deforche
Department of Public Health, Ghent University,
Ghent, Belgium
e-mail: jelle.vancauwenberg@ugent.be

B. Deforche
e-mail: benedicte.Deforche@UGent.be

J. Van Cauwenberg
Research Foundation Flanders (FWO),
Brussels, Belgium

© The Author(s) 2018

A. Curl and C. Musselwhite (eds.), *Geographies of Transport and Ageing*,
https://doi.org/10.1007/978-3-319-76360-6_6

(Sallis et al. 2004; Centers for Disease Control and Prevention 2013; Eurobarometer 2010). The vast majority of previous studies on active transport among older adults have focused on walking. Walking is the most accessible transportation mode, but cycling enables greater distances to be covered than walking and therefore may entail greater benefits for older adults' mobility and expand their activity space (Mandl et al. 2012; Patterson and Farber 2015). In this chapter, we provide an overview of studies examining the health benefits of cycling for transport, prevalence of cycling, determinants of cycling and bicycle crashes among older adults. Finally, we discuss the potential of electrically assisted bicycles (e-bikes) to stimulate cycling among older adults. It should be noted that cycling can be performed for different purposes, i.e. for transport (e.g. to a grocery store and a senior organization) and recreational purposes (e.g. cycling with friends along the sea side). These two types of cycling may entail different benefits and have different determinants. In this chapter, we focus on cycling for transport. Whenever research findings specifically about cycling for transport were unavailable, findings about overall cycling were presented.

Health Benefits of Cycling for Transport

In the general adult population, there is strong evidence that engagement in total cycling (i.e. for transport and recreation) has positive effects on cardiovascular fitness, cardiovascular and cancer risk and all-cause and cancer mortality (Oja et al. 2011). In a meta-analysis, it was shown that cycling 2.5 hours/week was associated with a 10% risk

---

B. de Geus · B. Deforche
Human Physiology Research Group, Department of Movement and Sport Sciences, Faculty of Physical Education and Physical Therapy, Vrije Universiteit Brussel, Brussels, Belgium
e-mail: bas.de.geus@vub.ac.be

B. Deforche
e-mail: benedicte.Deforche@UGent.be

reduction in all-cause mortality (Kelly et al. 2014). Focusing on older adults, a cross-sectional study in a Japanese rural area reported higher frequencies of cycling for transport to be related to higher levels of physical activity, a stronger social network and better mental health (Tsunoda et al. 2015). The latter two relationships may be explained by cycling increasing one's activity space resulting in more independence, the possibility to engage in social activities and maintain social relationships. Another cross-sectional study among urban Japanese older adults reported higher levels of cycling to be associated with a lower risk of limitations in instrumental activities of daily living (i.e. use public transportation independently and shop for daily necessities) and better social health among older adults who suffer from some mobility limitations (i.e. difficulties walking or climbing stairs) (Sakurai et al. 2016). A Dutch longitudinal study among 2109 men and women aged 55–85 years showed that cycling at least ten minutes/day at baseline was associated with smaller decreases in functional performance over three years compared to no cycling at baseline (Visser et al. 2002). A cross-sectional study and a pre-post cycling programme evaluation among middle-aged and older Australians indicated that cycling may result in balance improvement, which is important to prevent falls (Rissel et al. 2013). Participants in the cycling programme also reported increased fitness and leg strength as well as social and mental health benefits. In an experimental study in the UK, 77 non-cycling older adults (aged 50–83 years) assigned to an intervention group were provided with a cycle training assessment/skills development programme and then asked to cycle outdoors for at least 30 minutes, three times a week for a period of two months (Jones et al. 2016). Intervention participants rode a conventional pedal bike or an e-bike. Compared to the control group of non-cyclists, spatial reasoning, executive function and mental and physical health significantly improved among conventional bikers as well as e-bikers.

Based on the strong evidence for the beneficial effects of cycling on chronic disease risk and mortality in younger age groups and the favourable relationships/effects on mental and social health and physical functioning observed among older adults, it can be concluded that cycling can contribute to healthy ageing.

Prevalence of Cycling for Transport

Prevalence of cycling for transport varies greatly from country to country. In the general population, cycling shares of daily trips range from approximately 1% in the USA, Australia and Canada to 26% in the Netherlands. Among older adults, cycling shares are 0.5% in the USA, 1% in the UK, 9% in Germany, 15% in Denmark and 23% in the Netherlands (Buehler and Pucher 2012). In the state of Victoria (Australia), adults aged 65–85 years made less than 1% of their trips by bicycle, and no cycling trips were recorded for older adults aged older than 85 years (O'Hern and Oxley 2015). Recent studies in Japan reported high levels of cycling for transport, with 33.4 and 12.0% of urban and rural older adults cycling daily, respectively (Tsunoda et al. 2015; Sakurai et al. 2016). These differences in cycling prevalence between countries may reflect differences in climate, topography, culture and cycling infrastructure.

The majority of older adults' trips in general and by bicycle are for shopping purposes (Boschmann and Brady 2013; Winters et al. 2015b; O'Hern and Oxley 2015). Many of the trips that older adults undertake by car are within distances feasible for cycling (Janssens et al. 2013; Boschmann and Brady 2013). For example, in Flanders (Belgium), 23.7% of older adults reported cycling for transport almost daily (Van Cauwenberg et al. 2012) and cycling accounted for 17.5% of kilometres travelled per day (Janssens et al. 2013). Compared to other countries, this prevalence of cycling and contribution to overall travel is substantial. However, it has also been shown among Flemish older adults that 47.5% of all their trips shorter than three kilometres were done by private motorized transport, which illustrates the potential to further promote cycling for transport (Janssens et al. 2013).

It can be concluded that cycling for transportation differs greatly across different countries around the world. Promoting cycling among older adults in countries where cycling levels are low may result in substantial mobility benefits as well as physical, mental and social health benefits. In countries with relatively high levels of cycling, there is still room for improvement by promoting the substitution of motorized transport with cycling for short distance travel.

Determinants of Cycling for Transport Among Older Adults

In order to stimulate cycling for transport among older adults, knowledge is needed about its influencing factors, the so-called determinants of cycling for transport (Baranowski et al. 1998). Based on socio-ecological models, determinants can be divided into individual and environmental determinants (Sallis et al. 2006).

Individual Determinants

At the individual level, knowledge of socio-demographic characteristics related to levels of older adults' cycling for transport can help to identify subgroups with low levels of cycling for transport. It is important to identify such subgroups in order to design campaigns targeting increases in cycling in those subgroups that are most in need. In a population-based sample of 48,879 Belgian older adults, it was found that the probability of daily cycling for transport was lower for older participants, women, those with tertiary education and those with more functional limitations compared to their counterparts (Van Cauwenberg et al. 2012). A lower educational background is generally a risk factor for lower levels of physical activity and adverse health outcomes (Bauman et al. 2012; World Health Organization 2010). Cycling for transport may be an accessible form of physical activity for older adults from a lower socio-economic background and may in this way contribute to narrow social inequalities in health. In a mixed-methods study among 191 older adults living in Vancouver (Canada), the quantitative part showed that those aged older than 80 years, those living alone, those without access to a car and those with more co-morbidities were less likely to cycle (Winters et al. 2015a). Additionally, gender, income, employment status and ethnicity were found to be unrelated to the probability of cycling. Both the findings in Belgium and Canada suggest that cycling levels decrease with increasing age and deteriorating health (which also correlate with marital status and access to a car), which is not surprising since cycling requires a certain level of muscular strength and balance.

The qualitative part of the Canadian mixed-methods study showed that having a history of cycling was the most important facilitator of cycling. A study in the UK using semi-structured biographical interviews with 236 middle-aged and older adults (+50 years) revealed three cycling trajectories: reluctant riders (stopped or substantially decreased their cycling in the past five years), resilient riders (consistently cycled or increased their cycling) and re-engaged riders (restarted cycling after a hiatus) (Jones et al. 2016). Motivators and barriers towards cycling in these three groups were also explored. The reluctant riders did not cycle because they did not enjoy it or they stopped cycling because of an increased sense of vulnerability in heavily trafficked situations, the onset of a health condition, retirement or the loss of others to cycle with. However, most of the reluctant riders did acknowledge the health benefits of cycling. Most of the resilient riders had a strong history of cycling (i.e. commuting), and former commuter cyclists adapted their cycling habits after retirement. Re-engaged cyclists had restarted cycling often after decades without cycling. Getting fit and maintaining health were the main motivators for these cyclists. Cycling re-uptake was also encouraged by a partner who cycled. Relocation to an area supportive of cycling was also mentioned as a motivating factor to restart cycling. Zander et al. (2013) conducted a qualitative study to explore the motivators and barriers towards cycling experienced by seventeen Australians aged 49–72 years participating in a cycling promotion programme. The main reason for wanting to cycle was exercise, and the main barrier was fear of cars and traffic. Other motivators were enjoyment (e.g. being outdoors and feeling of freedom), social reasons (to spend time with family or make new friends) and the belief that cycling was easy to integrate into their daily lives. Furthermore, they perceived cycling to be a form of exercise that they can continue to do now that they are older. Participants also liked to use their bike for transportation because they considered it to be convenient, a good way to be active, and they enjoyed it and/or believed it had environmental benefits. Many of the participants did not feel confident in their riding ability since they had not ridden a bike in years. However, they did not seem to be particularly concerned about falling of their bikes. Based on their findings, the authors recommended future cycling programmes for older adults to include a skill course to develop cycling skills and built confidence, to provide local cycle maps with traffic-calm routes and a mentor (a local cyclist) to provide tips and support (Zander et al. 2013; Rissel et al. 2013). Several of these recommendations are related to environmental determinants of cycling, which are discussed in the next section.

Environmental Determinants

Several qualitative studies have indicated that safety is a major concern for cycling among older adults (Winters et al. 2015a; Van Cauwenberg et al. 2018; Jones et al. 2016). Qualitative interviews with older Canadian cyclists as well as non-cyclists revealed safety-related concerns to be an important barrier for cycling (Winters et al. 2015a). These concerns involved safety from motor vehicle traffic, behaviour of other cyclists and pedestrians as well as bicycle theft. Participants disliked sharing the road with motorized traffic and preferred traffic-calm roads or separated cycling paths. Besides fear from motorized traffic, they were also concerned about other cyclists and pedestrians not obeying the rules of road and creating dangerous situations on cycling paths. A study among Belgian older adults using bike-along interviews showed that traffic safety was their most important concern. The older adults felt safer when cycling on tracks that were well separated from motorized traffic and when crossings were clearly demarcated (i.e. regulated by traffic lights or indicated by markings) (Van Cauwenberg et al. 2018). The fear of cycling alongside motorized traffic also emerged from mobile observations and video elicitation interviews with 95 older cyclists (who had been cycling consistently in the last five years) in the UK (Jones et al. 2016). Although these could be considered experienced cyclists, they perceived cycling alongside motorized traffic to be dangerous, especially when riding alongside buses and heavy goods vehicles. Hence, participants liked to cycle on cycle paths, which they sometimes had to share with pedestrians. While they generally preferred biking on such paths, they also mentioned they have to be careful with pedestrians who behave less predictably, such as children and dog walkers. A recent systematic review summarizing the evidence on age variations in preferences for cycling provision separated from motor traffic showed that all age groups had a preference for separation compared to shared use with motorized traffic (Aldred et al. 2017). Some evidence suggested that older age groups may have a stronger preference for separation than younger age groups.

Poor-quality and narrow cycling paths and uneven surfacing were other factors that hindered older adults' cycling (Jones et al. 2016; Van Cauwenberg et al. 2018). Poor-quality surfacing was deemed to be a fall hazard, but also distracted participants from dangers posed by motorized traffic and diminished the possibility to enjoy the surroundings. Another issue discussed by older cyclists was poor legibility of traffic situations; this made them feel anxious, frustrated and vulnerable. In the Belgian study, participants mentioned making detours to avoid busy streets, dangerous crossings, inadequate cycling infrastructure and steep slopes (Van Cauwenberg et al. 2018). Hence, some degree of connectivity enabled participants to choose among alternative routes and made it more attractive to cycle for transport.

According to a systematic review by Cerin et al. (2017), only two studies previously examined the quantitative relationships between built environmental attributes and older adults' cycling for transport. Both studies were conducted in Belgium. In a first study, a lower probability of daily cycling for transport was observed in urban compared to semi-urban areas (Van Cauwenberg et al. 2012). The perceived presence of more shops and better access to public transport were related to higher probabilities of daily cycling for transport. Unexpectedly, lower levels of perceived traffic safety and the presence of litter and noise were found to relate to higher probabilities of daily cycling for transport (Van Cauwenberg et al. 2012). This may be attributable to important cycling destinations such as shops being located in areas with more motorized traffic and human activity, which may be associated with lower levels of perceived traffic safety and more litter and noise. Hence, some older adults need to cycle in such cycle-unfriendly environments to reach their destinations. The second study reported neighbourhood walkability, an index combining objectively assessed residential density, land use mix diversity and street connectivity, to be unrelated to cycling for transport (Van Holle et al. 2014).

Interactions between individual and environmental characteristics are a central premise of socio-ecological models (Stokols 1996). This implies that certain environmental characteristics will influence cycling among some subgroups of the older population, but not among other subgroups. For example, it has been suggested that women have a stronger risk aversion and that, therefore, the provision of cycling infrastructure that is well separated from motorized traffic may be particularly relevant to promote cycling among women (Garrard et al. 2008). In line with this, Van Cauwenberg et al. (2012) observed the presence of street lighting to be positively related to the odds of daily cycling for transport among female but not among male subgroups of Flemish older adults. However, no clear moderation effects of gender were observed for the other environmental factors studied. Relationships between environmental characteristics and cycling may also be moderated by psychosocial characteristics (e.g. attitudes towards cycling). For example, there is some evidence that environmental characteristics may be particularly relevant to stimulate active transport among adults with a general preference for passive transport (Van Dyck et al. 2009). Future studies should examine whether certain environmental characteristics are more or less important for certain subgroups in older adults (e.g. based on gender, socio-economic status, functional capacity and psychosocial characteristics). Such information would enable policy makers and planners to tailor environmental modifications to the needs of the most vulnerable subgroups or subgroups known to be at risk for low levels of cycling.

Studying bicycling commuting to work among 16- to 74-year-olds in the UK, Aldred and colleagues reported that local areas experiencing increases in bicycle commuting between 2001 and 2011 also experienced a decrease in the representation of older bicycle commuters (55–74 years) (Aldred et al. 2016). In other words, policy initiatives managed to successfully promote bicycle commuting among younger age groups, but not among the oldest age groups. These findings indicate that knowledge is needed about the determinants that are particularly important for older adults' cycling for transport and that these should be targeted by policy initiatives. To effectively promote cycling for transport among older adults, more research about its individual and environmental determinants and the interactions between them in different geographical and cultural contexts is necessary.

Bicycle Crashes Among Older Adults

In their Global Status Report on Road Safety, the World Health Organization (WHO) recognizes that road traffic deaths among cyclists are 'intolerably high' (World Health Organization 2015). For example in the Netherlands, more than half of all serious road injuries in 2011 were the result of a bicycle crash. Moreover, the share of seriously injured bicyclists has increased from 42% in 2000 to 60% in 2011 (Weijermars et al. 2016). While there is little research about older adults' cycling crashes, the studies conducted indicate that older cyclists appear to be particularly vulnerable (Mindell et al. 2012). A literature review of studies examining bicycle safety while taking into account exposure data (e.g. distance, duration or frequency of cycling) concluded that adults aged 50 years and older have a higher risk of being involved in a crash (Vanparijs et al. 2015). Furthermore, using data from the injury registration in a Swedish university hospital, Scheiman and colleagues found that older cyclists were more likely to suffer from more severe injuries (52%) than older car drivers (32%) and young cyclists (34%) (Scheiman et al. 2010). In both the USA and Germany, fatality and severe injury rates per 100 million kilometres cycled (defined as 'incidence rate' or 'risk') have been shown to be much higher than average for older cyclists, compared to middle-aged cyclists (Buehler and Pucher 2017). Furthermore, fatality and severe injury rates among older cyclists were, respectively, almost twice and five times as high in the USA compared to Germany. According to the authors, this difference may be partly attributable to the more extensive and better quality cycling infrastructure in Germany compared to the USA.

Sukarai and colleagues examined whether particular subgroups based on demographics and functionality had a higher probability of bicycle-related falls among 399 Japanese older adults (Sakurai et al. 2016). They observed a higher probability of falls among men, those with lower hand grip strength, a more depressed mood and those experiencing difficulties walking or climbing stairs. Age, BMI, cognitive functioning, frequency of going outdoors and frequency of cycling were not related to the probability of bicycle-related falls.

Single-Bicycle Crashes

Scheiman et al. (2010) reported that falling when getting on or off a bicycle was an important cause of all injuries (20%), especially among women. This type of incident was responsible for 43% of all hip and femur fractures, which are injuries with a heavy burden on the health care sector. Another common cause of injury among older cyclists was falling due to poor road surface quality (i.e. uneven or slippery surfaces, 26%). Only 6% of all injuries in older cyclists were caused by collisions with cars, buses or trucks. In the Netherlands, more than 80% of the seriously injured cyclists in between 2000 and 2011 were involved in a so-called single-bicycle crash or crashes between bicycles (Reurings et al. 2012; Schepers and Klein Wolt 2012). Schepers and Klein Wolt (2012) showed that older cyclists are more likely to sustain severe injuries in single-bicycle crashes. In another study, they showed that for 1996–2014, after controlling for age and taking into account exposure (kilometres cycled), the average annual cyclist deaths in crashes with motor vehicles decreased with 5.4% whereas the average annual deaths in crashes without motor vehicles increased with 4.4%. These trends were more favourable than those without controlling for age which can be explained by the older participants having an increased exposure (more kilometres cycled) and an increased risk (Schepers et al. 2017). It should be noted that, in most countries, road safety measures are based on police records. This seems to be a serious problem because studies showed that the police more often record crashes when a motorized vehicle is involved, resulting in an underestimation of the actual number of bicycle crashes and fatalities (De Geus et al. 2012).

Boele-Vos and colleagues conducted an in-depth study on the causes and circumstances of single-bicycle crashes and bicycle-slow traffic crashes involving cyclists aged 50 and over (Boele-Vos et al. 2017). The study was conducted in the Netherlands among 41 participants, cycling on different types of bicycles. Cyclists aged 75 and over seemed to be more frequently involved in falls from a bicycle compared to the 50–64 year age group. Bicycle crashes not involving high-speed traffic were divided into three types of crashes: (1) falls from a bicycle (when mounting or dismounting), (2) cyclists who collide with an object and (3) crashes between a bicycle and another road user who travels at a low speed (i.e. pedestrian, cyclist, moped rider or light-moped rider). Two-third of the crashes occurred on urban roads. About half of the urban crashes occurred in a 30 km/h speed limit zone, and the other half occurred in a 50 km/h speed limit zone. Forty-nine per cent of all crashes occurred on a cycle track (bicycle path or bicycle lane). The behaviour of another road user and too narrow bicycle facilities or carriageways were the most frequently described contributory factors to the crash. Other important crash contributors were distraction, narrow focus (focus on only one element of the traffic situation) and unusual traffic situations such as roadworks. Visibility of critical information in the visual periphery is important for safe cycling, especially for older adults (Schepers and den Brinker 2011). This population often has lower visual capabilities and decreased steering performance as a result of slower reaction times and decreased arm strength. Poorly visible kerbs, bollards and road narrowing cause many falls, especially under conditions with poor luminance (e.g. at night, rainy days or the absence of street lighting).

E-bikes: Potential Contributors to Older Adults' Mobility?

Electrically assisted bicycles (also called pedelecs or e-bikes), which are battery-driven and equipped with a torque or velocity sensor that triggers supporting power only when the cyclist exerts power onto the pedals, offer an alternative to 'regular' bicycles enabling people to cycle similar or greater distances with less physical effort (Fishman and Cherry 2016; Sperlich et al. 2012). For example among Dutch adults, it has been shown that the mean commuting distance with an e-bike is 1.5 times longer than with a conventional bicycle (Hendriksen et al. 2008).

Average intensity levels during e-biking substantially surpass three metabolic equivalents (METs), the minimum needed to be classified as moderately intensive and health-enhancing physical activity (Sperlich et al. 2012; Gojanovic et al. 2011). Furthermore, e-biking has been shown to result in an increase in physical fitness in middle-aged adults (De Geus et al. 2013). In a study among eight physically inactive women with a mean age of 38 years, total energy expenditure when cycling on an e-bike was 37% lower and mean power output was 29% lower compared to cycling on a conventional bicycle (Sperlich et al. 2012). The lower physical effort compared to regular cycling makes e-bikes especially attractive for older adults and may help to overcome longer distances and hilly and windy conditions. E-bikes' popularity has strongly increased during the past years (Papoutsi et al. 2014). In 2015, over 40 million e-bikes were sold worldwide and almost 37 million of these were sold in China (Fishman and Cherry 2016). In Europe, annual e-bike sales increased from 588,000 in 2010 to 1,357,000 in 2015 (CONEBI 2016). While the share of e-bikes compared to the total number of bikes (conventional bikes and e-bikes) sold remained low in some countries (e.g. 1.1% in Great Britain and 3.3% in France), it reached 11.0, 21.8 and 23.4% in Germany, the Netherlands and Belgium, respectively. In Flanders (the northern part of Belgium), 11.3% of all trips in the general population is made by bike and 0.8% by e-bike (Declercq et al. 2016). Among Flemish older adults, these figures are 8.8 and 3.0%, respectively. These figures show that (1) e-bikes are used particularly by older adults and (2) about one-quarter of all bike trips by Flemish older adults are made by e-bike. A market study by TNO in the Netherlands showed that 65% of e-bike owners were older than 65 years (Hendriksen et al. 2008).

Safety Aspects of E-bikes

While e-bikes may have a potential to contribute to older adults' mobility, their higher speeds and weights may constitute a risk for crashes and injuries. To our knowledge, only two studies previously examined crashes with e-bikes (Schepers et al. 2014; Papoutsi et al. 2014).

Schepers and colleagues undertook a case–control study to compare the likelihood of collisions for which treatment at a hospital emergency department was needed. Injury consequences were analysed for e-bike against conventional bicycle users aged 16 years or older in the Netherlands (Schepers et al. 2014). After controlling for age, gender and amount of bicycle use (exposure), e-bike users were found to be more likely to be involved in a crash requiring treatment at an emergency department. Crashes with e-bikes were about as equally severe as collisions with conventional bicycles, but older adults were at higher risk for (severe) injuries resulting from e-bike crashes. In Switzerland, Papoutsi and colleagues showed that only 17.4% of the hospitalized e-bike users were injured after a collision with a vehicle. Other injuries were the result of single-bicycle accidents (Papoutsi et al. 2014). Despite the popularity of e-bikes in some parts of the world, research about e-bike use, its potential contribution to older adults' mobility and potential safety issues is still in its infancy (Fishman and Cherry 2016).

Conclusions

The promotion of cycling among older adults entails important benefits for society and individuals. Safety appears to be a critical component of strategies aimed at promoting cycling among older adults. Increasing perceptions of safety appears to be key to stimulate cycling among older adults, and increasing actual levels of safety is needed to reduce the burden of cycling-related injuries. E-bikes may lower some of the barriers towards cycling and have the potential to contribute to older adults' mobility. In general, research about cycling among older adults is limited in comparison with 18- to 64-year-olds. To be able to utilize (e-) bikes' full potential to foster active and healthy ageing, more research about its benefits, determinants, injury risks and effective strategies to increase safe (e-)biking among older adults is necessary.

References

Aldred, R., J. Woodcock, and A. Goodman. 2016. Does More Cycling Mean More Diversity in Cycling? *Transport Reviews* 36: 28–44.

Aldred, R., B. Elliott, J. Woodcock, and A. Goodman. 2017. Cycling Provision Separated from Motor Traffic: A Systematic Review Exploring Whether Stated Preferences Vary by Gender and Age. *Transport Reviews* 37: 29–55.

Austroads. 2010. *The Australian National Cycling Strategy 2011–2016* [Online]. Sydney. Available at: http://www.bicycleinfo.nsw.gov.au/downloads/australian_national_cycling_strategy.pdf.

Baranowski, T., C. Anderson, and C. Carmack. 1998. Mediating Variable Framework in Physical Activity Interventions—How Are We Doing? How Might We Do Better? *American Journal of Preventive Medicine* 15: 266–297.

Bauman, A.E., R.S. Reis, J.F. Sallis, J.C. Wells, R.J.F. Loos, and B.W. Martin. 2012. Correlates of Physical Activity: Why Are Some People Physically Active and Others Not? *The Lancet, Physical Activity Series*, 31–44.

Biton, A., D. Daddio, and J. Andrew. 2014. *Statewide Pedestrian and Bicycle Planning Handbook*. Cambridge, MA: U.S. Department of Transportation.

Boele-Vos, M.J., K. Van Duijvenvoorde, M.J.A. Doumen, C. Duivenvoorden, W.J.R. Louwerse, and R.J. Davidse. 2017. Crashes Involving Cyclists Aged 50 and Over in the Netherlands: An In-depth Study. *Accident Analysis and Prevention* 105: 4–10.

Boschmann, E.E., and S.A. Brady. 2013. Travel Behaviors, Sustainable Mobility, and Transit-Oriented Developments: A Travel Counts Analysis of Older Adults in the Denver, Colorado Metropolitan Area. *Journal of Transport Geography* 33: 1–11.

Buehler, R., and J. Pucher. 2012. Walking and Cycling in Western Europe and the United States. *TR News* 280: 34–42.

Buehler, R., and J. Pucher. 2017. Trends in Walking and Cycling Safety: Recent Evidence from High-Income Countries, with a Focus on the United States and Germany. *American Journal of Public Health* 107: 281–287.

Centers for Disease Control and Prevention. 2013. *U.S. Physical Activity Statistics*.

Cerin, E., A. Nathan, J. Van Cauwenberg, D. Barnett, and A. Barnett. 2017. The Neighbourhood Physical Environment and Active Travel in Older Adults: A Systematic Review and Meta-Analysis. *International Journal of Behavioral Nutrition and Physical Activity* 14.

CONEBI. 2016. *European Bicycle Market 2016 Edition—Industry & Market Profile (2015 Statistics)*. Brussels, Belgium: Confederation of the European Bicycle Industry.

Declercq, K., S. Reumers, E. Polders, D. Janssens, and G. Wets. 2016. Onderzoek Verplaatsingsgedrag Vlaanderen 5.1 (2015–2016): Tabellenrapport. Instituut voor Mobiliteit (Universiteit Hasselt, in opdracht van de Vlaamse Overheid).

De Geus, B., F. Kempenaers, P. Lataire, and R. Meeusen. 2013. Influence of Electrically Assisted Cycling on Physiological Parameters in Untrained Subjects. *European Journal of Sport Science* 13: 290–294.

De Geus, B., G. Vandenbulcke, L. Int Panis, I. Thomas, B. Degraeuwe, E. Cumps, J. Aertsens, R. Torfs, and R. Meeusen. 2012. A Prospective Cohort Study on Minor Accidents Involving Commuter Cyclists in Belgium. *Accident Analysis and Prevention* 45: 683–693.

Eurobarometer. 2010. *Sport and Physical Activity*. Available at: http://ec.europa.eu/public_opinion/archives/ebs/ebs_334_en.pdf.

European Parliament. 2010. *The Promotion of Cycling*. Brussels: Policy Department B: Structural and Cohesion Policies.

Fishman, E., and C. Cherry. 2016. E-bikes in the Mainstream: Reviewing a Decade of Research. *Transport Reviews* 36: 72–91.

Garrard, J., G. Rose, and S.K. Lo. 2008. Promoting Transportation Cycling for Women: The Role of Bicycle Infrastructure. *Preventive Medicine* 46: 55–59.

Gojanovic, B., J. Welker, K. Iglesias, C. Daucourt, and G. Gremion. 2011. Electric Bicycles as a New Active Transportation Modality to Promote Health. *Medicine and Science in Sports and Exercise* 43: 2204–2210.

Hendriksen, I., L. Engbers, J. Schrijver, R. Van Gijlswijk, J. Weltevreden, and J. Wilting. 2008. *Elektrisch Fietsen. Marktonderzoek en verkenning toekomstmogelijkheden*. TNO Kwaliteit van Leven.

Janssens, D., K. Declercq, and G. Wets. 2013. *Onderzoek Verplaatsingsgedrag Vlaanderen 4.4 (2011–2012)*. Instituut voor Mobiliteit.

Jones, T., K. Chatterjee, J. Spinney, E. Street, C. Van Reekum, B. Spencer, H. Jones, L. Leyland, C. Mann, S. Williams, and N. Beale. 2016. *Cycle BOOM Design for Lifelong Health and Wellbeing—Summary of Key Findings and Recommendations*. UK: Oxford Brookes University.

Kelly, P., S. Kahlmeier, T. Goetschi, N. Orsini, J. Richards, N. Roberts, P. Scarborough, and C. Foster. 2014. Systematic Review and Meta-Analysis of Reduction in All-Cause Mortality from Walking and Cycling and Shape of Dose Response Relationship. *International Journal of Behavioral Nutrition and Physical Activity* 11.

Mandl, B., A. Millonig, S. Klettner, M. McDonald, N. Hounsell, A. Wong, B. Shrestha, N. Baldanzini, A. Penumaka, and I. Hendriksen. 2012. Growing Older, Staying Mobile: Transport Needs for an Ageing Society. Deliverable D4.2 Older People Walking and Cycling. GOAL Consortium.

Mindell, J.S., D. Leslie, and M. Wardlaw. 2012. Exposure-Based, 'Like-for-Like' Assessment of Road Safety by Travel Mode Using Routine Health Data. *PLoS ONE* 7: e50606.

O'Hern, S., and J. Oxley. 2015. Understanding Travel Patterns to Support Safe Active Transport for Older Adults. *Journal of Transport & Health* 2: 79–85.

Oja, P., S. Titze, A. Bauman, B. de Geus, P. Krenn, B. Reger-Nash, and T. Kohlberger. 2011. Health Benefits of Cycling: A Systematic Review. *Scandinavian Journal of Medicine and Science in Sports* 21: 496–509.

Papoutsi, S., L. Martinolli, C.T. Braun, and A.K. Exadaktylos. 2014. E-bike Injuries: Experience from an Urban Emergency Department-a Retrospective Study from Switzerland. *Emergency Medicine International* 2014: 850236.

Patterson, Z., and S. Farber. 2015. Potential Path Areas and Activity Spaces in Application: A Review. *Transport Reviews* 35: 679–700.

Reurings, M.C.B., W.P. Vlakveld, D.A.M. Twisk, A. Dijkstra, and W. Wijnen. 2012. Van fietsongeval naar maatregelen: Kennis en hiaten; Inventarisatie ten behoevevan de nationale onderzoeksagenda fietsveiligheid (NOAF) [From Bicycle Crashes to Measures: Knowledge and Knowledge Gaps; Inventory For the Benefit of the National Research Agenda Bicycle Safety (NOAF)]. In *Research*, ed. S. I. F. R. S. The Hague.

Rissel, C., E. Passmore, C. Mason, and D. Merom. 2013. Two Pilot Studies of the Effect of Bicycling on Balance and Leg Strength Among Older Adults. *Journal of Environmental & Public Health* 2013: 1–6.

Sakurai, R., H. Kawai, H. Yoshida, T. Fukaya, H. Suzuki, H. Kim, H. Hirano, K. Ihara, S. Obuchi, and Y. Fujiwara. 2016. Can You Ride a Bicycle? The Ability to Ride a Bicycle Prevents Reduced Social Function in Older Adults with Mobility Limitation. *Journal of Epidemiology* 26: 307–314.

Sallis, J., L. Frank, B. Saelens, and M. Kraft. 2004. Active Transportation and Physical Activity: Opportunities for Collaboration on Transportation and Public Opportunities Health Research. *Transportation Research Part a-Policy and Practice* 38: 249–268.

Sallis, J., R. Cervero, W. Ascher, K. Henderson, M. Kraft, and J. Kerr. 2006. An Ecological Approach to Creating Active Living Communities. *Annual Review of Public Health* 27: 297–322.

Scheiman, S., H.S. Moghaddas, U. Bjornstig, P.O. Bylund, and B.I. Saveman. 2010. Bicycle Injury Events Among Older Adults in Northern Sweden: A 10-Year Population Based Study. *Accident Analysis and Prevention* 42: 758–763.

Schepers, P., and B. den Brinker. 2011. What Do Cyclists Need to See to Avoid Single-Bicycle Crashes? *Ergonomics* 54: 315–327.

Schepers, P., and K. Klein Wolt. 2012. Single-Bicycle Crash Types and Characteristics. *Cycling Research International* 2: 119–135.

Schepers, P., E. Fishman, P. den Hertog, K. Wolt, and A. Schwab. 2014. The Safety of Electrically Assisted Bicycles Compared to Classic Bicycles. *Accident Analysis and Prevention* 73: 174–180.

Schepers, P., H. Stipdonk, R. Methorst, and J. Olivier. 2017. Bicycle Fatalities: Trends in Crashes with and Without Motor Vehicles in the Netherlands. *Transportation Research Part F* 46: 491–499.

Sperlich, B., C. Zinner, K. Hebert-Losier, D.P. Born, and H.C. Holmberg. 2012. Biomechanical, Cardiorespiratory, Metabolic and Perceived Responses to Electrically Assisted Cycling. *European Journal of Applied Physiology* 112: 4015–4025.

Stokols, D. 1996. Translating Social Ecological Theory into Guidelines for Community Health Promotion. *American Journal of Health Promotion* 10: 282–298.

Thiry, C. 2011. IRIS2. Brussels: Mobiel Brussel-BUV van het Brussels Hoofdstedelijk Gewest.

Tsunoda, K., N. Kitano, Y. Kai, T. Tsuji, Y. Soma, T. Jindo, J. Yoon, and T. Okura. 2015. Transportation Mode Usage and Physical, Mental and Social Functions in Older Japanese Adults. *Journal of Transport & Health* 2: 44–49.

Van Cauwenberg, J., P. Clarys, I. De Bourdeaudhuij, V. Van Holle, D. Verte, N. De Witte, L. De Donder, T. Buffel, S. Dury, and B. Deforche. 2012. Physical Environmental Factors Related to Walking and Cycling in Older Adults: The Belgian Aging Studies. *BMC Public Health* 12.

Van Cauwenberg, J., P. Clarys, I. De Bourdeaudhuij, A. Ghekiere, B. De Geus, N. Owen, and B. Deforche. 2018. Environmental Influences on Older Adults' Transportation Cycling Experiences: A Study Using Bike-Along Interviews. *Landscape and Urban Planning* 169: 37–46.

Van Dyck, D., B. Deforche, G. Cardon, and I. De Bourdeaudhuij. 2009. Neighbourhood Walkability and Its Particular Importance for Adults with a Preference for Passive Transport. *Health & Place* 15: 496–504.

Van Holle, V., J. Van Cauwenberg, D. Van Dyck, B. Deforche, N. Van de Weghe, and I. De Bourdeaudhuij. 2014. Relationship Between Neighborhood Walkability and Older Adults' Physical Activity: Results from the Belgian Environmental Physical Activity Study in Seniors (BEPAS Seniors). *International Journal of Behavioral Nutrition and Physical Activity* 11.

Vanparijs, J., L.I. Panis, R. Meeusen, and B. de Geus. 2015. Exposure Measurement in Bicycle Safety Analysis: A Review of the Literature. *Accident Analysis and Prevention* 84: 9–19.

Visser, M., S.M.F. Pluijm, V.S. Stel, R.J. Bosscher, and D.J.H. Deeg. 2002. Physical Activity as a Determinant of Change in Mobility Performance: The Longitudinal Aging Study Amsterdam. *Journal of the American Geriatrics Society* 50: 1774–1781.

Weijermars, W., N. Bos, and H.L. Stipdonk. 2016. Serious Road Injuries in the Netherlands Dissected. *Traffic Injury Prevention* 17: 73–79.

Winters, M., J. Sims-Gould, T. Franke, and H. McKay. 2015a. "I Grew Up on a Bike": Cycling and Older Adults. *Journal of Transport & Health* 2: 58–67.

Winters, M., C. Voss, M.C. Ashe, K. Gutteridge, H. McKay, and J. Sims-Gould. 2015b. Where Do They Go and How Do They Get There? Older Adults' Travel Behaviour in a Highly Walkable Environment. *Social Science and Medicine* 133: 304–312.

World Health Organization. 2010. *Equity, Social Determinants and Public Health Programmes*. In ed. E. Blas, and A.S. Kurup. Geneva, Switzerland: World Health Organization.

World Health Organization. 2015. *Global Status Report on Road Safety 2015*. Geneva, Switzerland: World Health Organization.

Zander, A., E. Passmore, C. Mason, and C. Rissel. 2013. Joy, Exercise, Enjoyment, Getting Out: A Qualitative Study of Older People's Experience of Cycling in Sydney, Australia. *Journal of Environmental and Public Health* 12: 1–6.

7

Out-of-Home Mobility of Senior Citizens in Kochi, India

Talat Munshi, Midhun Sankar and Dhruvi Kothari

Background

The increase in the number of seniors[1] living in Indian cities and the challenges they face is becoming a serious concern. In 2014, more than 100 million Indians were considered seniors, estimates suggest that by

[1]Seniors are citizens above 60 years of age, and they have also been referred to as seniors or aged in other literature.

T. Munshi (✉)
UNEP DTU Partnership, Technical University of Denmark, Copenhagen, Denmark
e-mail: tamu@dtu.dk

M. Sankar
Urban Mass Transit Company (UMTC), Gurgaon, Haryana, India
e-mail: midhun.sankar.k@gmail.com

D. Kothari
University of Pennsylvania, Philadelphia, USA
e-mail: msdakot@gmail.com

© The Author(s) 2018

A. Curl and C. Musselwhite (eds.), *Geographies of Transport and Ageing*,
https://doi.org/10.1007/978-3-319-76360-6_7

2050 around 250–300 million Indian (more than 1 in 10) will be seniors (Subaiya and Bansod 2014). A large portion of these will reside in urban areas and are expected to live an independent life, as the demographic structure of India is moving from a joint family structure (a unified large family) to a nuclear family concept of "me my wife and my children" (Patel 2005). Increased life expectancy in India (Riley 2001) would mean that a large portion of these seniors will be in the higher age brackets.

Given these demographic transitions, it is important that Indian cities develop such that they can offer a supportive environment, enabling residents to grow older and remain active within their families, neighbourhoods and civil society. Overall, the current urban and transport system in India remains very poor (Pucher et al. 2005; Munshi et al. 2013; Munshi 2016) and its provision is largely unresponsive to the needs of seniors (Munshi et al. 2004). Therefore, seniors face problems related to mobility and accessibility to their basic needs (Chokkanathan and Lee 2006). A typical Indian city is characterized by rapid growth, limited and outdated transport infrastructure, suburban sprawl, sharply rising motor vehicle ownership and use, deteriorating public transport services, practically no footpaths and bicycle lanes, and inadequate as well as uncoordinated land use and transport planning (Munshi et al. 2014). Thus, in a condition where the mainstream population itself experiences a crisis of inadequate supply, the mobility challenges for seniors can be extreme. These mobility restrictions have an adverse effect on the quality of life of older adults by threatening independent living and personal autonomy.

This paper looks at mobility habits of seniors in the town of Kochi in the state of Kerala, India, and studies how the household living arrangement influences the outdoor mobility of seniors. Kochi was considered as a case, as the city is growing rapidly and is the largest city in the state of Kerala.[2] People of Kerala State who emigrate to other countries for work find it comfortable to settle in an urban set up like Kochi when they return back to Kerala. As a result of this trend, the city of Kochi is

[2]Kerala also has a higher, that is, 12% of its population who are senior citizens compared to the national share of 8%.

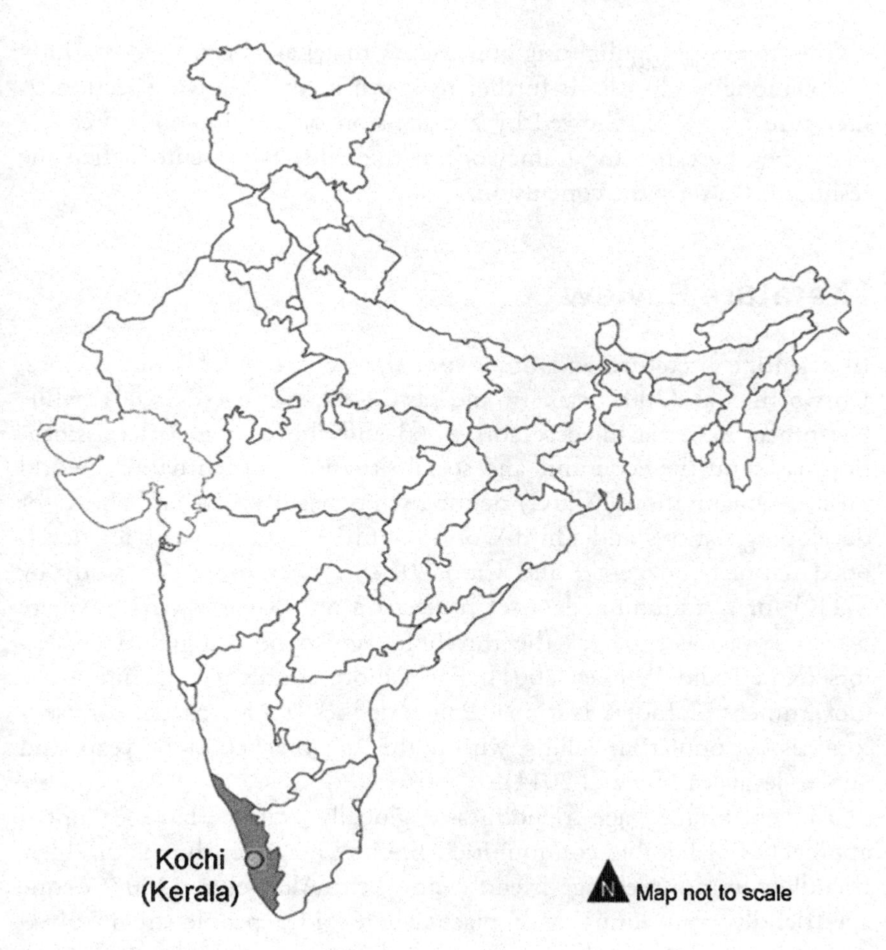

Fig. 7.1 Location map—Kochi

the preferred location to settle for seniors and has the highest number of old-age homes in Kerala (Fig. 7.1).

Kochi is the largest urban agglomeration in Kerala and the city has typical infrastructure and mobility problems associated with a similarly sized city in India. The results from the study are specific to the city of Kochi in Kerala but lessons learnt are relevant to other similar cities in India.

The literature highlighting approaches that have been used to study senior-mobility globally is further reviewed in the following section. A literature review is followed by a discussion on the methods used for this paper, hereafter the framework of the study is presented, then the results and latterly the conclusion.

Literature Review

In scientific literature and otherwise, the definition of seniors varies. Unsworth et al. (2001) classify "old" as 65–80 and "aged" as 80+, while few others have classified seniors as 63 and above. A global classification based on the economic and social situation of each nation, World Health Organisation (WHO) defined seniors as 60 years and above for developing nations and Third-World countries, and 65 years for developed nations (Fitzgerald and Caro 2014). An example is a study by WHO for a minimum data set project on Sub-Saharan Africa, where 50 years was accepted as the threshold age to be recognized as seniors (Kowal and Peachey 2001). For a more contextual definition the Government of India's National Policy defines "senior citizen" or "seniors" as a population falling within the age bracket of 60 years and above (Jeyalakshmi et al. 2011).

To encourage age-friendliness, globally cities have adopted approaches of liveable communities, lifetime neighbourhoods or elder-friendly communities, age-friendly cities, etc. Alley et al. (2007) define age-friendly community as a place where older people are involved actively, respected and supported with infrastructure and services (Fitzgerald and Caro 2014). Active ageing and mobility are closely related, thus access to needs and transport infrastructure support are key age-friendly support elements.

For this type of planning and design practices across the world, two features focussing on physical environment and social environment are widely accepted (Lui et al. 2009). Literature has proven that social relationships and activity interactions are important elements to quality of life for seniors, but with increasing age, these elements become more difficult to maintain. Consequently, mobility becomes fundamental for seniors' overall well-being and active participation. Outdoor mobility is often referred to as the ability to move about—either ambulant, using an assistive device or by the means of transportation facilities—sufficiently to carry out outside-home activities. Such outdoor mobility becomes a prerequisite, not only for obtaining essential day-to-day commodities and consumer goods but also for general well-being. Therefore, transport and mobility find mention across all concepts along with housing and built environment.

Webber et al. (2010) categorized senior-mobility as a composition of five fundamental determinants, that is, cognitive, psychosocial, physical, environmental (Iwarsson and Ståhl 2003), and financial; while gender, culture, and biography (personal life history) have cross-cutting influence over these fundamental determinants of senior-mobility. Apart from physical and financial abilities which are frequently scripted, cognitive health issues such as mental status, memory and reaction time affect the ability of seniors to remain mobile and active. Additionally, psychological factors like self-efficacy, depression, fear and relationships with others have an impact on the belief seniors have about their own self-efficacy. This becomes critical in determining the extent of seniors' mobility and access to society. In general, it has been found that older people are likely to travel less compared to younger people, the number of journeys declines with age, and the trips get shorter (Metz 2000). It is also found that in developed nations, private automobile dependence increases with age (Scheiner 2006). But with age, seniors also prefer to travel as passengers due to eyesight, confidence to drive, health condition, etc., making them rethink the choice of self-driving (Fiedler 2007) and less likely to choose public transport modes (Rosenbloom 2005). Even if public transit is physically accessible to them, older adults have a variety of safety, personal, security, flexibility, reliability and comfort concerns (Rosenbloom 2005). The Activity of Daily Living approach (Lawton 1971) assesses the ability to carry out instrumental or/and physical activities (Peel et al. 2005; Baker et al. 2003). The life space assessment approach measures the frequency of travel and assesses the need for assistance. It helps in identifying mobility at all location levels categorized as the bedroom, home, outside home, neighbourhood, town and unlimited space. Carp (1988) in his congruence model conceptualized the elder mobility as a function of the degree of individual's fitness, needs and the environment. He categorized health care, food and clothing as life maintenance needs and recreational activities and social relationships as higher order needs. The fulfilment of the life maintenance need is linked to independent living and the higher order need is linked to the well-being of a senior individual. Fulfilment of needs is thus an important aspect of the senior's life, in which mobility plays a crucial role.

Approach and Methods

This study focuses on finding the "Influence of household living arrangement on the outdoor mobility of seniors", which concentrates on (1) the activities and needs of seniors, (2) the travel behaviour of seniors, and (3) the effect of household living arrangement on the mobility of seniors. Given the context of the economic and social setting, this study will consider population above 60 years as seniors.

To understand the activities and need of seniors, their activities are classified according to Carp's Model with some modifications to suit the context. The activities are categorized into (1) life-maintenance needs—daily survival needs (work, shop, medical and services), and (2) higher order needs—needs resulting in well-being (recreation, social, religious and others). The categorization helps in understanding and generalizing the necessity of needs, which is an important factor in senior-mobility. The travel behaviour of seniors is studied by analysing travel needs, modes used, time of travel, frequency, dependency, mobility options available, etc. To understand the influence of the household living arrangement, a classification of seniors according to their living arrangement was made. However, cases of living arrangement in this study are limited to the people living with offspring and independently/senior couples.

To understand the various domains of influence on the seniors' life, it was logical to go for a constructive approach, as it tends to use open-ended questions so that the participants can share their views

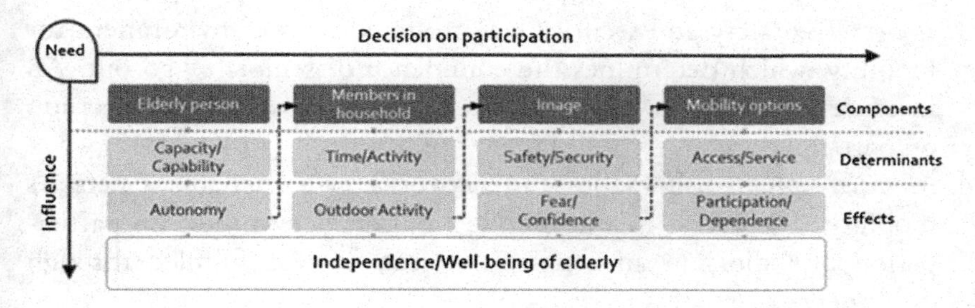

Fig. 7.2 Frame for the study

(Creswell 2014). A pilot survey was conducted to understand specific contexts, which included unstructured interviews and expert interview for validation of the initial framework. The framework of the study was later restructured to suit the context to further understand the relationship with and influence of household living arrangement on the mobility of seniors. A thorough analysis was done to categorize many key words used in the pilot study. Figure 7.2 is the framework used in this study to understand the relationship of the household living arrangement with the mobility of seniors.

The framework explains the decision-making process of seniors while participating in an activity. The framework was constructed with an understanding that various factors influence senior-mobility. The framework has four verticals, which are the four components in the senior-mobility decision-making process. The framework also has three horizontals for each component which influences the senior-mobility. The working of the framework is explained below.

1. *Seniors*: The capacity (health condition) and capability (financial, social condition) that determine autonomy in the decision on participation in an activity resulting in outdoor mobility.
2. *Member living in the household*: Members such as spouse/children/other members generate activities, which might involve senior's participation such as taking care of kids and homemaking if members are working. These activities determine the time allotment for the outdoor activity of seniors.

3. *Image*: The safety and security factor offered by the environment for mobility, which determines the confidence of seniors to go out. An insecure environment might have a negative impact on the decision on participation.
4. *Mobility options*: All modes and networks, which enable a person's mobility. The access to the mobility option determines the participation of seniors in an activity that can be also fulfilled through dependence.

If a senior were able to access the mobility option and fulfil his or her needs, this would indicate the well-being and the independence of the seniors.

Semi-structured interviews, telephone interviews and focus group discussions were conducted with a sample of 89 seniors, which were selected through convenience sampling. An attempt was made to capture a range of people with different capabilities, capacities, social backgrounds, etc. Distribution of the sample is shown in Table 7.1.

Table 7.1 Sample descriptive

Capability/classes	Female				Male				Total
	60–70	70–80	80+	Total	60–70	70–80	80+	Total	
No. of people surveyed	29	6	4	39	24	21	5	50	89
Mobility impairments									
Perfectly alright	24	3	1	28	19	11	2	32	60
Can walk, but tiring	5	2	1	8	5	8	2	15	23
Equipment's help[a]	–	1	1	2	–	2	1	3	5
Other person's help[b]	–	–	1	1	–	–	–	–	1
Hearing impairments									
Can hear perfectly	28	5	–	33	24	19	3	46	79
Hearing impairments[c]	1	1	4	6	–	2	2	4	10

[a]Individual who can move with assistance from mobility aids and disability equipment
[b]Individual who can only move when assisted by others
[c]Individual who finds hearing a challenge

Thematic analysis was conducted on the data collected through semi-structured interviews (recordings, notes, images, observation, etc.) which were arranged and prepared for analysis in terms of narratives. Key words were categorized into various domains such as types of needs, modes used, frequency, relations, activity and dependency. These were further used to highlight the influence of the household living arrangement on the senior-mobility by finding relations with other domains.

Results

This study gives an insight into seniors' experience of limited mobility in Kochi. Some of the limitations observed were dependency, accessibility, social stigma of being aged, etc., living arrangement being the key influencing factor in the fulfilment of their needs.

Seniors were mostly found to be dependent on their children or others to satisfy most of their daily needs and activities. It has been observed that limited autonomy associated with ageing tends to give a general feeling of increased helplessness and seclusion. In terms of mobility parameters such as trips, seniors preferred short trips, most of which were to places that they were familiar with. A majority of seniors were reluctant to travel when the distance of the trip was longer. Household living conditions and image of a particular location had a very strong influence on how and where the seniors travelled. The study, however, tried to understand the influencing factors according to the framework set, which has been elaborated in the following section.

Seniors Person's Capability and Capacity

In the interviews conducted, many referred to old age as a *terrible phase of life* and suggested people should be prepared for it. Ageing reduces the capability and the capacity of individuals and affects their activities and need to travel. Due to diminished capabilities, seniors tend to travel less. Further, increased traffic and poor transport infrastructure were also found to inhibit travel. Seniors in Kochi mostly avoided using self-owned motorized modes for travel; only a very few from those interviewed reported to have used motorized modes. Because seniors found it difficult to use motorized modes and bicycles, most were captive walkers, not as a preference but due to compulsion. Most seniors reported that increased traffic congestion and poor pedestrian facilities restrict their capacity to walk. This is supported by the statement from the seventy-five-year-old male (Box 1) who said that people like him own cars, but find it difficult to drive and manoeuvre in chaotic traffic conditions. He also highlighted that it is also difficult to find a parking space near the destination, forcing him to walk for a significant distance to the destination. He also associates this with his diminishing capabilities due to ageing. The seniors also recalled and compared their old days and present days; one senior who recently injured his hip stated, "I have cycled through almost the whole city, but now I am confined to this house". The reluctance to travel was echoed by a senior man (81) and another female (71), they felt that at this age "they should not be going to meet people; rather others should come and meet them".

Box 1: Capabilities

...if I take an auto I can get down to the place where I want to...now due to being aged I find difficult to drive through this busy traffic...

—Male, 75

Household Living Arrangements

Having limited capacity to move and with very little support (infrastructure and otherwise) systems, the household condition and living arrangements (seniors, seniors couple living together and seniors living with offspring) affected how seniors viewed their basic needs and their accessibility to these. Seniors living with their children felt the comfort of being taken care of. However, living arrangements and dependency also forced seniors to move and leave their children, some seniors cited having physiological effects, which was evident from a statement of a 78-year-old woman's interview "Even though she and he (daughter and son in law) takes care of me well, I just don't feel very comfortable there. I like our home". The role they assume or are assigned within the household can restrict the individual's ability, as it affects the amount of time they have to participate in other activities. The family situation also induces coupling constraints (Hägerstrand 1970) as individuals have to coordinate their activities with other members of the family. The statements in Box 2 exemplify the stated influences. In the first case, the senior woman is actively involved in homemaking and has very little free time and independence of her own.

Box 2: Activities

...I am mostly busy home making, taking care of children... both of them (son & daughter in law) are working now... Sometimes on weekends we go out...

—Female, 65, living with children

...I just have to prepare for two...How much time will we both sit in the house and watch TV? So evenings we go for a walk to park...

—Female, 72, living with spouse

In the case of seniors living alone or with their spouse, outdoor trips were relatively higher and so was participation in recreational, religious, social activities as they had more time in hand. As citied by the second female in Box 2, two seniors take care of the homemaking, but as the number of members in the house is less, a change in the activity pattern was still possible. This allowed them the time and space to participate in outdoor activities. However, they still felt constrained in satisfying their daily needs. Senior women living alone said "I buy things for a month in one go" and as a senior male living with offspring saying, "he (son) buys things when he visits us while returning from his office". They also had a feeling of boredom as they had more time on hand and restricted mobility forcing them to stay indoors. Individuals living as a senior couple (separate from children) want to live close to activities that satisfy their basic needs, for example health facilities and religious visits.

The death of a spouse affects seniors' mobility in both positive and negative ways. Two statements in Box 3 explain how the death of a family member affected the seniors' activity. The death of spouse socially restricts the senior women from making outdoor trips, especially social and recreational ones as part of local customs and rituals. Most of the senior women said they would strictly following these customs.

Box 3: Intangible affects

... after her (wife's) death, I actually stopped going to park and all...

—Male, 72, living with children

...I shifted to kid's place after his (spouse's) death... now I go to many places with them (son and family)...

—Male, 72, living with children

Among all the basic needs, as can be expected, seniors were most concerned about their health and medical needs. Seniors who live alone or with their spouses have a feeling of insecurity in case of an emergency (health or any other). They fear that in case of medical emergency, they might not be able to reach the nearest health facility. It was clear from some statements that seniors made reference to access to health facilities such as "Yes, that is a problem" while another female stated "it is always good to have some people you know around you, when there is a medical case or check-up". Seniors living alone created a good network of people who they could rely on in the absence of their immediate family members, such as trusted drivers, neighbours to tackle situations.

Religious trips remained a high-priority trip for all seniors irrespective of their living arrangement. For many, religious trips were the only outdoor activity they performed. Religious beliefs and devotion have a high role in the seniors' well-being as the religious places were also a place for social interaction with their peer group, which made seniors enthusiastic about religious trips. Most of the seniors believed that they should be spiritual and it is an essential part of their life. Box 4 compares two cases of different living arrangements, which shows the significance of religious beliefs in the seniors' life. In one case, one senior male living with his spouse, whom were abandoned by their children, feels helpless and feels god is the only motivation in his life. While in the second case, the senior female shifted to the new flat bought by her son. She is happy about the move, but complains about not being able to visit the temple regularly. These cases show the importance of religious trips for seniors, which has lesser direct relation with the seniors' living arrangement. The study found that household living arrangement has less influence on the participation of older people in religious trips.

Box 4: Religious

...In this phase of life, you only have god...

—Male, 78, living with spouse

...Our earlier house was very near to the temple, after shifting to this flat now going to temple has become difficult...

—Female, 67, living with children

Mobility Image

The mobility image of a particular location can instil fear in seniors, and it is possible that they lose confidence to travel to a particular location. Conversely, a good experience can also instil confidence in seniors to travel again to the same location. This image is a result of their own experience or a result of experience narrated by someone else (in person or via a media like newspapers). Most respondents reported bad travel experiences and in fact, almost all had experienced an accident or a near accident-like situation in the past year. This instilled fear in them to travel, and in general, they avoided travelling to locations with excessive traffic because of these experiences. Many seniors also reported being abused on the street, which was another factor that they reported as one restraining their travel. As said by a senior male "the way auto (auto rickshaws) drivers behave… it's humiliating". Box 5 has two examples of seniors in different living arrangements facing a similar problem. A senior woman who lives alone changed her mode preference to auto rickshaw after being in an accident, which has a higher travel cost associated with it. Where as in case 2, a senior male explains his mode choice for his work trip after his experience of an accident, but he relies on his son to drop him at his shop. Many seniors were also resigned to the situation. They have lost hope, and most cannot imagine that the road and traffic conditions could be better and more friendly to them. As said by a senior "I wouldn't blame the road, after all, they are meant for vehicles".

Box 5: Image & Mode choice

...even though it is expensive, I take auto, after that accident ...

—Female, 73, Living alone

...I would previously walk to my shop, but after I met with an accident I rely on my son to drop me to shop...

—Male, 68, living with children

The fear instilled due the mobility image forced seniors to be dependent, as their children also discourage seniors to travel by their own because of the fear. In this case, the autonomy, as well as the independence of seniors, is affected.

Mobility Options

For autonomy of the seniors, accessibility to available mobility options is important, especially to public transport options. Seniors faced problems while boarding a bus and while crossing roads. A senior in an interview explained his experience on travelling in the city buses "They just don't wait before you board they drive off". Seniors also mentioned that speeding of buses, and sudden acceleration and deceleration of the bus, made it very difficult for them to use it. It was also found that most seniors, especially in the higher age bracket, were averse to using technology like the direct-to-home delivery system, where food and groceries get delivered to their home. Some of them also reported having ordered cabs using a mobile phone application, but most found it too complicated to operate and some were unable to comprehend the system. Seniors also found dependency handy when it came to making longer trips. As stated in Box 6, a senior male presents difficulties in arranging a driver for long-distance travels, where as in the second case, the female very casually discusses the trip. It also shows the ease of depending on one's children.

Box 6: Mobility Options

...we mostly avoid social functions which happen at our native place, we have to arrange a driver for such long drives...

—Male, 76, living with spouse

...yeah we go along with them (children), all we have to do is just dress and sit in the car right (she laughs off) ...

—Female, 61, living with children

Currently, the mobility situation for seniors seems like a downward spiral. Unless systemic intervention is put in place, which ensures that seniors at least have good access to basic necessities and are able to live independent and respectful lives, things are likely to get worse. There are good examples from the West, some of which have been cited earlier, which cities like Kochi need to learn from. At the least, they should start with the provision of pedestrian infrastructure so that it is easy for seniors to walk and be mobile. Of course, it is important that the government and society, in general, think about their health and social needs and provide infrastructure for it.

Conclusion

This chapter contributes to the ongoing discussion about the necessity for active seniors living in cities. For Kochi city in Kerala, India, issues and constraints faced by seniors are discussed in relation to: the different mobility options they have, their own capacity and capability, their living condition and their perceptions. In Kochi, provision and design of transport infrastructure and the urban environment presents significant mobility barriers to the seniors restricting their accessibility to religious and healthcare facilities, and other essential basic services. Disproportionate growth of private automobiles and poor implementation of parking and driving regulation have contributed towards an image of roads being unsafe in the minds of seniors, fear of which inhibits them from travelling. This leads to a situation where the well-being of seniors is largely dependent on support from their siblings and other well-wishers. If past trends continue, the number of senior residents is likely to increase dramatically as well as the number of private automobiles. The consequence of these on the demand and provision of mobility services for seniors (which need to complement the provision of mobility infrastructure) will be significant. If nothing is done, the situation is likely to get worse. However, it is unethical that people find it difficult to meet their basic needs and are unable to access healthcare and religious activities.

For seniors to live an active life, clear, inclusive and senior-friendly urban design and stress on infrastructure for walking and public transport are necessary to improve access to their daily needs, places of worship and locations where they can socialize. The National Urban Transport Policy (NUTP) (MOUD 2006a) and the proposed Green Urban Transport Scheme (MOUD 2006b) are measures taken by Indian Government in the correct direction. These should trickle down and help the Municipal Government of Kochi to provide walking and public transport infrastructure. There are examples of inclusive planning methods (Mahadevia et al. 2014) which Kochi can use to plan inclusive and seniors-friendly urban environments. The next generation of seniors will find it is easier to use information technology; therefore, affordable on-call mobility services can also be encouraged to provide mobility to seniors. The key to improving mobility of seniors is providing more mode choices and options. With changing demographic trends, it is in the best interest of society that appropriately planned mobility options are provided in such a manner that seniors can live respectful lives even when they choose to live independently.

References

Alley, D., P. Liebig, J. Pynoos, T. Banerjee, and I.H. Choi. 2007. Creating Elder-Friendly Communities: Preparations for an Aging Society. *Journal of Gerontological Social Work* 49: 1–18.

Baker, P.S., E.V. Bodner, and R.M. Allman. 2003. Measuring Life-Space Mobility in Community-Dwelling Older Adults. *Journal of the American Geriatrics Society* 51: 1610–1614.

Carp, F.M. 1988. Significance of Mobility for the Well-Being of the Elderly. *Transportation in an Aging Society: Improving Mobility and Safety of Older Persons* 2: 1–20.

Chokkanathan, S., and A.E. Lee. 2006. Elder Mistreatment in Urban India: A Community Based Study. *Journal of Elder Abuse & Neglect* 17: 45–61.

Creswell, J.W. 2014. *Research Design: Qualitative, Quantitative, and Mixed Methods Approaches*. London: Sage.

Fiedler, M. 2007. Older People and Public Transport: Challenges and Changes of an Ageing Society. ETMA (European Metropolitan Transport Authorities) Rupprecht Consult.

Fitzgerald, K.G., and F.G. Caro. 2014. An Overview of Age-Friendly Cities and Communities Around the World. *Journal of Aging & Social Policy* 26: 1–18.

Hägerstrand, T. 1970. What About People in Regional Science? *Papers in Regional Science* 24: 6–21.

Iwarsson, S., and A. Ståhl. 2003. Accessibility, Usability and Universal Design—Positioning and Definition of Concepts Describing Person-Environment Relationships. *Disability and Rehabilitation* 25: 57–66.

Jeyalakshmi, S., S. Chakrabarti, and N. Gupta. 2011. Situation Analysis of the Elderly in India. *Central Statistics Office, Ministry of Statistics and Programme Implementation, Government of India Document.*

Kowal, P., and K. Peachey. 2001. Indicators for the Minimum Data Set Project on Ageing: A Critical Review in Sub-Saharan Africa. *Dar es Salaam, United Republic of Tanzania: WHO*, 9.

Lawton, M.P. 1971. The Functional Assessment of Elderly People. *Journal of the American Geriatrics Society* 19: 465–481.

Lui, C.W., J.A. Everingham, J. Warburton, M. Cuthill, and H. Bartlett. 2009. What Makes a Community Age-Friendly: A Review of International Literature. *Australasian Journal on Ageing* 28: 116–121.

Mahadevia, D., T. Munshi, R. Joshi, K. Shah, Y. Joseph, and D. Advani. 2014. *A Methodology for Local Accessibility Planning in Indian Cities*. Ahmedabad: Centre for Urban Equity, CEPT University.

Metz, D.H. 2000. Mobility of Older People and Their Quality of Life. *Transport Policy* 7: 149–152.

MOUD. 2006a. *National Urban Transport Policy*. India: Government of India.

MOUD. 2006b. *New Green Urban Transport Scheme* [Online]. http://pib.nic.in/newsite/PrintRelease.aspx?relid=153393. Accessed 23 August 2017.

Munshi, T. 2016. Built Environment and Mode Choice Relationship for Commute Travel in the City of Rajkot, India. *Transportation Research Part D: Transport and Environment* 44: 239–253.
Munshi, T., W. Belal, and M. Dijst. 2004. Public Transport Provision in Ahmedabad, India: Accessibility to Work Place. In *Urban Transport X: Urban Transport and the Environment in the 21st Century*, ed. L.C. Wadhwa. Southampton: WIT Press.
Munshi, T., M.F.A.M.V. Maarseveen, and M.H.P. Zuidgeest. 2013. *Built Form, Travel Behaviour and Low Carbon Development in Ahmedabad, India*. PhD, University of Twente.
Munshi, T., K. Shah, A. Vaid, V. Sharma, K. Joy, S. Roy, D. Advani, and Y. Joseph. 2014. Low Carbon Comprehensive Mobility Plan, Rajkot. Unpublished Report, UNEP Risoe Centre on Energy, Climate and Sustainable Development, Technical University of Denmark, Roskilde.
Patel, T. 2005. *The Family in India: Structure and Practice*. New Delhi: Sage.
Peel, C., P.S. Baker, D.L. Roth, and C.J. Brown. 2005. Assessing Mobility in Older Adults: The UAB Study of Aging Life-Space Assessment. *Physical Therapy* 85: 1008.
Pucher, J., N. Korattyswaropam, N. Mittal, and N. Ittyerah. 2005. Urban Transport Crisis in India. *Transport Policy* 12: 185–198.
Riley, J.C. 2001. *Rising Life Expectancy: A Global History*. Cambridge: Cambridge University Press.
Rosenbloom, S. 2005. The Mobility Needs of Older Americans. In *Taking the High Road: A Transportation Agenda for Strengthening Metropolitan Areas*, ed. B. Katz and R. Puentes. Washington, DC: Brookings Press.
Scheiner, J. 2006. Does the Car Make Elderly People Happy and Mobile? Settlement Structures, Car Availability and Leisure Mobility of the Elderly. *European Journal of Transport and Infrastructure Research* 6: 151–172.
Subaiya, L., and D.W. Bansod. 2014. Demographics of Population Ageing in India. In *Population Ageing in India*, ed. G. Girdhar et al., 1–41. Delhi: Cambridge University Press.
Unsworth, K.L., K.J. McKee, and C. Mulligan. 2001. When Does Old Age Begin? The Role of Attitudes in Age Parameter Placement. *Social Psychology Review* 3: 5–15.
Webber, S.C., M.M. Porter, and V.H. Menec. 2010. Mobility in Older Adults: A Comprehensive Framework. *The Gerontologist* 50 (4): 443–450.

8

Walking with Older Adults as a Geographical Method

Angela Curl, Sara Tilley and Jelle Van Cauwenberg

Introduction

For most people, walking is an everyday practice. Walking can be performed for recreation or transportation purposes and is a function of the individual and their interaction with social and built environment. For older adults, the built environment can play a greater role in the ability to walk outdoors compared with younger populations and walking itself can become a more important mode of transport in older age, particularly due to driving cessation. Walking is something that many older people can do and is therefore

A. Curl (✉)
Department of Geography, University of Canterbury—Te Whare Wānanga O Waitaha, Christchurch, New Zealand
e-mail: angela.curl@canterbury.ac.nz

S. Tilley
OPENSpace Research Centre, Edinburgh College of Art, University of Edinburgh, Edinburgh, UK
e-mail: sara.tilley@ed.ac.uk

© The Author(s) 2018
A. Curl and C. Musselwhite (eds.), *Geographies of Transport and Ageing*,
https://doi.org/10.1007/978-3-319-76360-6_8

important for physical activity. Engaging with older adults as they walk allows us to research how and why people walk and barriers to doing so.

This chapter is about walking with older adults as a method for researching walking practices. Walking interviews are introduced as an example of go-along or mobile methods, which are particularly useful as an approach to understanding older adults' experiences of and interactions with the outdoor environment. The chapter draws on the authors' experiences of go-along interviews with older adults in three different projects in Belgium and Scotland. The chapter starts with a background to go-along interviews as a geographical method and the relevance for mobilities research, particularly with older adults. Following the background, we present three case studies based on our fieldwork experiences. The purpose of the case studies is to demonstrate different research contexts for studies utilising walking interviews and to discuss the practicalities associated with conducting go-along interviews with older people. The third section is a reflection on the process of go-along interviews. We discuss the benefits and limitations of this approach related to broader issues in research interviews including power relations, ethics and risk. The fourth section focusses on practical advice and considerations for those planning go-along interviews with older adults.

Background and Literature

Mobility is important for health, well-being and quality of life (Gatrell 2013) for populations of all ages, but has particular pertinence for older adults who may experience changes in functional ability and more

J. Van Cauwenberg
Department of Public Health, Ghent University,
Ghent, Belgium
e-mail: jelle.vancauwenberg@ugent.be

J. Van Cauwenberg
Research Foundation Flanders (FWO),
Brussels, Belgium

immediate health effects related to physical inactivity or immobility. Walking can be considered as the most basic form of mobility since it requires no specialised equipment nor specialised skills (compared with car driving or cycling). Therefore, walking can be described as an ideal form of physical activity (Morris and Hardman 1997) to promote among older adults, 60–70% of whom are insufficiently physically active (Centers for Disease Control and Prevention 2013; European Commission 2013). The outdoor environment interacts with individual characteristics to determine walking across spatial scales, encompassing access to amenities at a city level through to micro-scale influences such as kerb height (Sallis et al. 2006).

Micro-scale influences may be particularly relevant for older adults' walking since functional limitations may mean that barriers such as high kerbs, uneven or slippery pavements or long crossing distances present a greater challenge to mobility. The influence of these micro-scale characteristics on mobility is difficult to capture quantitatively since they are complicated to measure and may vary substantially within an older adult's neighbourhood, although there have been attempts to audit outdoor environments at the micro-level for walkability (Millington et al. 2009), including some which focus on older adults (Curl et al. 2016; Michael et al. 2009). Since micro-scale characteristics are environmental details and often context-specific, they may be hard to recall and report during indoor interviews. During walking interviews, older adults are exposed to the micro-scale environmental factors, which can trigger a discussion on how they influence walking experiences.

Walking is an inherently mobile practice. Researching while *on the move* allows consideration of how walking is enacted and the interactions between people and place which inform exactly how, where and when walking does or does not occur. Go-along interviews are an example of a mobile method and have been described as an 'imaginative' approach (Gatrell 2013) to understanding interactions between people and place, in contrast to quantitative studies of built environment impacts on health. Mobile methods have emerged alongside and in response to the new mobilities paradigm. Cummins et al. (2007) outline how place can be understood from a relational perspective and following this perspective go-along interviews focus on mobile individuals, layers of assets, dynamic characteristics, infrastructure and meaning, and variable descriptions of contextual features. Rather than separating contextual or environmental effects on the walking behaviour of older adults from the compositional effects, use of go alongs allows the processes and interactions between older adults and the places they walk to be explored. A mobilities perspective allows a focus on context, and by using mobile methods in mobilities research, we can begin to understand the importance of place and context on older adults' experiences of being mobile in the outdoor environment. Merriman (2014) has cautioned against over-reliance on mobile methods for mobilities research. However, when researching how people negotiate the built environment and the interactions between infrastructure, spaces and movement we posit that mobile methods are useful not only for allowing the researcher to experience with the participant, but also that by being in place, research participants are better placed to reflect on their own interactions and relations with the outdoor environment than when out of context. Such methods also provide the researcher with first hand exposure to others' abilities and perceptions of the environment (Cook et al. 2016). Being in the outdoor environment can stimulate memory and places can provide stimulus (Cook et al. 2016) for discussion. That is not to say mobile methods or the use of go-along interviews are a panacea or can 'reveal the truth' of mobility (Cook et al. 2016). In fact, when researching older adults' mobilities, it is particularly important to recognise that experiences of those who are less mobile and do not spend time outdoors are also crucial. Use of a go-along interview implies the ability to go along, and the limitations of this approach should be recognised by those doing this type of research with older adults.

In this section, we review studies which have used go-along interviews to study walking and relationships with the built environment. These studies do not necessarily have a specific focus on older adults. Butler and Derrett performed walk-along interviews with four disabled persons (aged 30–60 years) who had recently suffered from an injury (Butler and Derrett 2003). The walk-along interviews were part of a large-scale multi-method prospective outcome of injury study in

New Zealand. Given that many older adults suffer from mobility limitations, this study's findings are relevant for older adults. The participants were asked to choose an everyday route they like to share with the researcher and to talk about what helped and hindered their walking after the injury. The authors concluded that 'walking lends itself to a particular way of chatting about the physical reality of staying upright in the world: it is easy to talk about gait and balance when they are being demonstrated' (p. 6). Furthermore, it was stated that the walk provided an ideal background to discuss how participants experienced their disability. In a paper by Jones et al. (2008), three case studies are presented to illustrate how qualitative data obtained from go-along interviews can be linked to particular places. In one of these case studies, the researchers aimed to explore urban design and distinctive types of buildings and spaces in a large redevelopment site in Bristol. Participants were allowed to choose their own route through the redevelopment site, and the routes were recorded by means of geographical positioning systems (GPS). This allowed matching of participants' quotes with specific site locations. Furthermore, it allowed creation of maps in geographic information systems (GIS) depicting the routes chosen by the participants, which allowed areas that were frequently chosen to walk around to be studied in more detail. In another paper, Jones et al. (2011) describe a study using go-along interviews along a predetermined route to examine issues related to fear of crime among 19- to 22-year-olds. During the go-along interviews, participants were asked to press one of four buttons on a hand-held computer to indicate how happy they felt in relation to their personal safety. The hand-held computer also recorded the GPS location where the buttons were being pressed. Combining the maps of all participants enabled researchers to get an overview of which areas were generally considered to be unsafe (also called fear 'hotspots'). These areas could then be examined in detail through the interview data.

All the above studies focussed on walking interviews, but mobile methods can also be used to examine other transport behaviours, for example cycling. In the cycle BOOM-project, mobile observations and video-elicitation interviews were used to examine 95 older adults' experiences while cycling in the UK (Jones et al. 2016). About half of the sample cycled along a route selected by the participant while the other half cycled along a predefined route that varied on physical characteristics (e.g. on-road cycle lanes and off-road cycle tracks). Both the participant and researcher were equipped with an action video camera and a GPS device. Some participants were also equipped with a galvanic skin response sensor (i.e. a bio-sensing device that measures skin conductance levels) to assess emotional arousal (i.e. stress and relaxation) along the routes. After the ride, a video-elicitation interview was performed based on the video recordings of the bike ride. The interview data were then synchronised with the GPS and galvanic skin response data. The data delivered detailed insights into when and where older adults cycle and how they handle situations that they perceive as unsafe. The qualitative information was supported by the quantitative galvanic skin response data. For example, raised levels of stress corresponded with video observations and participant narratives about unpredictable pedestrian activity on shared paths. The cycle BOOM-project did not interview their participants during the bike ride itself, possibly for practical and safety reasons. In a Belgian project, 40 older adults cycled together with the researcher to a randomly chosen destination within their neighbourhood (Van Cauwenberg et al. 2018). The participants were equipped with a sports camera and were asked to describe out loud all environmental factors influencing their cycling for transport while cycling to the destination. The researcher asked questions whenever possible (e.g. in a traffic-calmed street or when waiting at a traffic light). After the ride, a video-elicitation interview based on the recordings was performed to complement the data collected during the bike-along interviews. A similar study was conducted among 35 Belgian primary schoolchildren (Ghekiere et al. 2014).

GPS are now carried by many people in the form of a smartphone. While, as other chapters in this book have identified, use of technology by older adults might be less prevalent than in the general population, use of smartphones is fairly and increasingly common. Combined with go-along interviews, the use of such technologies offers great potential for contextualising qualitative data using mobile methods. The use of GPS tracking allows a greater understanding of the relationships between individuals, mobility and place that has been impossible until relatively recently. For example, Cummins et al. (2007) suggest that tracing individuals throughout a period of time would allow a greater understanding of the impact of place on health outcomes, beyond traditional exposure assessments. Walking can in some ways be considered such a health outcome. Jones and Evans (2012) discuss the idea of a 'spatial transcript' and use of GPS in qualitative research. One option would be to review the route and any photographs or transcripts with participants after a walk. Combined use of GIS/GPS and qualitative methods allows analysis of diverse experiences, perspectives and subjectivities into studies measuring aspects of the built environment and offers some potential to address what has been called the quantitative—qualitative divide in Transport Geography (Goetz et al. 2009).

Walking with Older Adults: Case Studies

In this section, we draw on three examples studies undertaken by the authors which demonstrate different uses of walking methods with older adults.

Case Study 1: Mobility, Mood and Place: Walking interviews to explore the older people's emotional responses to place in Edinburgh, Scotland

The Mobility, Mood and Place (MMP) project[1] explored how places can be designed to make pedestrian mobility easy, enjoyable and meaningful into older age. Walking interviews were used to gain a deeper understanding of older people's emotional interactions and responses with the routes that they walked as such methods allow for richer accounts of perceptions of the environment than conventional, static surveys (Kusenbach 2003; Miaux et al. 2010).

Purposive sampling was used to recruit 19 diverse older people (defined as being aged 65 and over). A diverse age range was sought, in addition to an even gender split. They were all able to walk for at least 15 minutes in the local urban environment unaided by another person. Participants were recruited from across Edinburgh in order to provide a variety of different walks in different local contexts with variations between environments. The majority of participants started the walk at

[1]https://sites.eca.ed.ac.uk/mmp/.

their residential home; however, in a few instances, participants were met at a different location, e.g. in the local park, to begin the walk.

Participants were asked to undertake a 'typical walking journey', defined by them, which they would make in a typical week, and was at least 15 minutes long. Walks ranged from 20 minutes to 2 hours and were highly varied. They included trips to the shops, strolls around the neighbourhood, journeys to the bus stop and visits to parks and other places of interest. These took place during the summer of 2015. The weather was dry for all of the interviews, apart from two.

The route was recorded using GPS to enable it to be plotted visually. Photographs were also taken along the route by the researcher to record aspects of the environment that participants discussed during the interview. Interviews were recorded using a digital voice recorder with a clip-on microphone, with a wind guard. Interviews were transcribed, and the transcripts were used in an inductive thematic analysis. The walks all took place within Edinburgh across three broad types of area. Six took place in the urban city centre, six were in green spaces such as parks and woodlands, and seven of the walks were in locations where blue space (e.g. rivers) was a prominent feature.

Three key themes emerged from the walking interviews: the importance of memories and familiarity; the attraction of colourful visual experiences; and the importance of social contact and interaction. These themes relate to interactions between emotions, mood and the environments and are important in mobility choices in the local environment. For example, older people sought out walking routes in familiar places that had strong memories, particularly those that were more positive and poignant. Some of these environments supported welcomed periods of reflection. Walking interviews helped to understand the significance of these places by observing participants in them.

Participants were asked to explain and rationalise their route choices. This helped the researcher understand the decision-making process when walking a given route and understanding how it might change based on the interactions with the environment during the walk.

Case Study 2: Walk-along interviews to explore the environmental factors influencing Belgian older adults' walking for transport

In this example, walk-along interviews were used to explore perceived environmental influences on older adults' walking for transport. The walk-along methodology was selected since there was limited information about the environmental factors influencing older adults' walking for transport, especially in Flanders (Northern, Dutch-speaking part of Belgium). In addition, we were interested in uncovering the detailed street characteristics that influenced older adults' experiences while walking for transport. The knowledge gained from the walk-along interviews served as a basis for a series of studies using photographs to examine how (changes in) street characteristics influenced a street's appeal for walking for transport among older adults (Van Cauwenberg et al. 2014a, b, 2016).

The walk-along interviews were conducted with 57 older adults recruited by means of convenience and snowball-sampling (i.e. relatives of the research team were recruited). Participants had to be community-dwelling and able to walk independently for at least 30 minutes. They resided in three different urban or semi-urban regions in Flanders to ensure that a variety of environments were encountered during the walk-along interviews. Participants were visited at home where the walk-along interview was initiated after administration of a questionnaire assessing socio-demographics, functional limitations and physical activity levels. The walk-along interview was performed during a walk from the participant's home to a randomly chosen destination (e.g. supermarket, bakery, church and post office) located within a 10–15-minute walk. The route to the destination was the route participants usually chose when walking to the destination and the route back from the destination was chosen by the researcher based on a map of the participant's neighbourhood. Doing so, participants were exposed to familiar and less familiar walking routes and to a wider variety of environmental stimuli. The walk-along interviews lasted approximately 30 minutes. Before starting the walk-along interview, participants were provided with standardised instructions asking them to describe all things in the environment that facilitated or hindered their walking for transport. During the development of these instructions, the authors searched for a balance between making clear what was expected from the participants and not emphasising particular environmental factors (which may influence participants' behaviours and answers during the interviews).

During the walk-along interviews, participants were prompted to discuss environmental aspects influencing their walking experience and the researchers asked for additional information or detail whenever necessary. When participants discussed an environmental barrier, suggestions for improvement were asked for. All environmental factors discussed during the walk-along interview were photographed to illustrate our findings. The whole interview was tape-recorded and transcribed verbatim afterwards. An inductive approach based on grounded theory was adopted to analyse the qualitative data. For more details about the methods applied during the walk-along interviews and analyses, readers are referred to Van Cauwenberg et al. (2012).

The environmental factors discussed were categorised into eight categories: access to facilities, walking facilities, traffic safety, familiarity, safety from crime, social contacts, aesthetics and weather. Within these categories, environmental factors were described in great detail and with reference to older adults' particular needs. For example, when participants discussed the importance of sidewalk width, they mentioned that a sidewalk should be wide enough to easily pass with a wheelchair (when accompanying a relative using a wheelchair). Participants mentioned specific environmental features that diminished the useable width of the sidewalk, such as construction works, parked cars, unkempt greenery and utility or light poles. When discussing the presence of safe crossings, participants reported deviating from the shortest route to be able to safely cross a street (i.e. in the presence of a zebra crossing or traffic light). Such context-specific information would have been difficult to gather without having participants exposed to the specific situations. Being in situ also prompted participants to consider different environmental factors simultaneously. For example, the following participant weighed the presence of hazardous traffic against having to walk along a hazardous walking path: 'There's a lot of traffic over here. Especially during the morning, then here's a long traffic congestion. I prefer a bad and uneven path without traffic compared to this even sidewalk with all this traffic. I always avoid the traffic as much as possible, even when the alternative paths are bad, muddy or whatever'.

Besides obtaining detailed and context-specific information, the authors experienced that the participants enjoyed the walk-along interviews; they liked to share their experiences about a topic that really mattered to them, and their enthusiastic contributions confirmed the importance of the research topic. For the authors, the walk-along interviews yielded important insights into which, how and why environmental factors influenced Flemish older adults' walking for transport, which would have been difficult to obtain using other research methodologies.

Case Study 3: Getting Outdoors, Falls, Ageing and Resilience (GoFAR) Project[2]

The aim of this research project was to understand which features in the built environment best support older people's confidence to go outside without falling in order to inform development of an audit tool to help occupational therapists identify aspects of the environment that have caused people to fall or fear falling. Go-along interviews were chosen as the most appropriate method for engaging with older adults who had

[2] http://www.salford.ac.uk/built-environment/research/research-centres/surface/research/going-outdoors-falls,-ageing-and-resilience-go-far.

fallen to develop an understanding of their experiences of the built environment and which aspects of the environment can lead to falling or fear of falling.

As detailed in Curl et al. (2016), 20 adults aged over 65 were interviewed in autumn 2012 in four locations across the UK: Swansea, Edinburgh, London and Salford. All participants had fallen over outdoors in the previous 12 months. Participants were recruited through a linked study, so all had already participated in a focus group discussing falling outdoors (Nyman et al. 2013). This was a convenience sample with the criteria being aged over 65 and having experienced a fall in the past 12 months. Walks covered on average a distance of 1.32 km (sd = 0.72), taking 25.1 minutes (sd = 10.92).

The weather was mild with temperatures ranging from 10 to 15 °C and some light rainfall during walks.

When undertaking walking interviews, the route the walks should take is a key consideration. Consideration was given to whether the walk should be in a familiar environment for each participant or whether each participant should take the same route for purposes of the detailed audit of a particular environment, reflecting the variation among participants' perceptions as well as validating items within a controlled environment. To allow participants to walk in a familiar environment and to highlight environmental features that they perceived to be beneficial or deterring. Given all of the participants had previously fallen, it was deemed preferable to undertake the research in environments familiar to the participants, given that unfamiliar environments can increase the risk of falls (Phillips et al. 2013). A predetermined route was prepared to relieve pressure if participants could not easily decide on a route but this was not required by any of the participants.

Our route selection approach followed the work of Van Cauwenberg et al. (2012) who used pre-information on participants' daily activities to help decide the destination and route, integrated with the personal projects approach (Little 1983; Wallenius 1999). Personal projects are used to elicit the kind of outdoor activities participants undertake and were used to determine the route of the walk by walking to one of the places listed. Participants completed a short questionnaire before the walk to collect demographic data and information about their day-to-day activities.

In addition to the pre-questionnaire, three kinds of data were collected during the walk. Interviews were audio recorded using a digital voice recorder. Photographs were taken of a key positive or negative environmental features discussed by participants. In this study, the interviewer took the photographs to remove additional burden or avoid creating distractions for participants. The photographs were geo-tagged so they could be matched up with interview and GPS data as described below.

Each walk was recorded using a mobile phone geo-tracking application. The purpose of this was to record the length and time of the walks but also to allow the matching of interview recordings using a timestamp. Photographs were also integrated with the GPS and interview recordings. Interview recordings, transcripts, GPS tracks and photographs were stored and analysed using NVivo which allows for analysis of multiple qualitative media.

Our study focussed on identifying environmental features related to falls as reported in Curl et al. (2016). However, identifying such features is difficult as the influence of the built environment on falling or fear of falling varies based on individual and temporal factors. The process of walking interviews in this case highlighted the highly complex and interrelated nature of a range of both built environmental and individual influences relating to falls. Environmental features, weather, social relations, time of day, health and other personal characteristics intersect. For example, participants would avoid certain routes depending on the time of day or the weather as they have a detailed understanding of where water accumulates on the pavement making it difficult to navigate if it has rained.

Being in situ is hugely beneficial for this type of research. Being in the environment prompts participants to recall important features which they may struggle to do in a static interview. Furthermore, it is clear that the environment plays a fundamental role in triggering memory and emotional responses.

Discussion and Reflection

This section presents key themes which emerge from drawing together our experiences of walking with older adults in the case studies outlined. There are clear benefits to walking as a research method with older adults, but there are also disadvantages. We reflect on these here before providing some practical recommendations for those who wish to use this method in the following section.

In all of our case studies, walking interviews allowed us as researchers to explore older adults' experiences of the built environment, in place. This meant the discussions were grounded in place and not talking about abstract issues which are hard for the participant to describe and the interviewer to relate to. For example, rather than describing stairs as generally problematic participants could explain and demonstrate why they struggled to negotiate particular environments. This richer interaction and depth of engagement with participants and place have also been noted by others (Evans and Jones 2011).

This engagement with place meant that meaning and emotions attached to places were raised as important in understanding participants' mobility in the urban environment. It became clear that it is not only quantifiable or measurable aspects of the built environment which influence mobility, but the way in which people feel and the history and memories attached to that space. For example, in the GoFAR project, one of the interviewees explained the difficulties in walking beyond a certain point. There was nothing outwardly noticeable about the space, but they explained this had been the location of a previous fall and therefore presented a barrier to mobility beyond this point. Similarly, in the Belgian study, participants described that they liked to walk in familiar areas (e.g. a neighbourhood where they used to live) since it provided them with senses of safety and nostalgia. This highlights the importance of understanding the individual experiences and perceptions of the environment when considering walking behaviour, beyond any attempts to measure or generalise. Being in place meant that memories were triggered. Participants remembered events that had happened in particular places. Such depth of discussion would have been harder in a static environment, where it might be difficult to probe and more challenging for participants to recall their experiences. What becomes clear from these walking interviews is the importance of the intersection between individuals, spaces and others in influencing the mobility of older adults in urban environments. As highlighted by Cummins et al. (2007), there is a need to move away from separation of contextual and compositional influences on (health) behaviours, such as walking, and recognise diverse experiences, perceptions and relationships between people and place. This can be seen as relational geography or person-environment fit in environmental psychology.

It is well known that *being outside* can have positive effects on cognitive function (de Vries 2010) and mental well-being—doing research outside allowed participants the freedom to think and express themselves in a more natural setting. Simply by going for a walk, the power relations which may constrain interviews in some cases are removed, and the 'interview' becomes a chat, guided by the participant in their familiar environment. While walking interviews have been described as empowering (Carpiano 2009), they are not immune to issues of power or decisions about where is a suitable location (or route) for an interview (Jones et al. 2008).

Walking also highlighted the importance of the social environment and the intersection of the built environment with individuals and their social networks. Some environments were difficult to negotiate on their own, but fine with someone else. Having social support could make the difference between being comfortable in an environment or not as highlighted in the MMP study where social interaction was a key theme.

One thing which struck us as researchers is the multiple strategies that those with mobility difficulties enact in order to remain mobile. Participants described in detail their strategies for negotiating difficult environments and how they use objects in often complex strategies for mobility. For example, one participant would use the bus to go to the shops, but walk home, which might seem counter-intuitive to outsiders, but once their shopping trolley was full, they were able to use it as counterbalance to avoid falling, where the empty trolley did not provide the same support.

The temporal component of interactions with the built environment was also clear and related to strategies employed by participants. Participants would go to certain places at certain times of day because they were more accessible at certain times for various reasons, related to daylight, busyness or noise. For example, the time of day was important in some cases because of the shadows created by trees and buildings which could be disorientating depending on the position of the sun. This highlights that the role of the built environment and that its influence on mobility is emphasised for some older adults and can vary temporally as well as spatially. Vallée (2017) has recently called for more attention to be paid to temporal influences of the built environment, called the 'daycourse of place'.

Limitations of Walking Interviews

Walking interviews are clearly beneficial for researching walking, particularly when the research is focussed, as ours was on relationships between the built environment and mobility. However, there are also a number of limitations or criticisms of walking interviews as a research method. First, a 'walking interview' implies walking. It therefore by its very nature might exclude those who cannot walk, who cannot walk far or who are not motivated or confident to walk. It is important to consider who might be excluded through non-participation and therefore whose views are not reflected. Perhaps 'go-along' is a better term as this can mean moving, by whatever means and so is a more inclusive term when considering that that movement may be aided by a wheelchair or mobility scooter. Nevertheless, in our samples of older adults, none of the participants were using such mobility aids, and therefore it is important to consider whether the research method is exclusionary.

In the MMP study, all participants self-identified as healthy adults, they generally did not face problems walking in the environment and the majority of them were keen walkers. Their experiences of the urban environment may be very different to those that are less able to walk, not keen or confident in walking. However, using a sample, that is experienced in walking, could help to understand why they continue to be mobile in older age.

It is important to be clear that in all of our case studies, our research focus was walking. Yet, walking interviews may also be useful when the focus is not walking. As we have noted, the outdoor environment may be seen as more neutral, removing power imbalances and may therefore facilitate research with harder to reach groups or research on sensitive topics. Walking alongside removes eye contact so people may feel more comfortable discussing difficult topics.

The ethical implications of undertaking research of this nature with people who in some cases may be particularly vulnerable should also be considered. Particularly in the case of the project on falls, where the sample criteria included having recently fallen over. Given that one of the biggest risk factors for a fall is having already fallen (Scheffer et al. 2008), this group were vulnerable to falling and therefore were at risk by participating in the research.[3] Distractions can also cause falls, and so it is important to be aware that the interview itself may be a distraction.

[3] The project was approved by the University of Edinburgh Human Ethics Committee.

Responsibilities of the researcher need to be considered in this context, ensuring that when walking with older adults for the purposes of research participants are in as natural a context as possible and do not, for example leave a mobile telephone at home, forget keys or go along an unfamiliar route because of the presence of a researcher. Some of these issues lead to the considerations and practical advice; we outline in the next section.

As mixed methods researchers, we all have experience working with a range of methods researching issues of mobility in the built environment and feel that walking interviews have been one of the most interesting and informative approaches. Participants were more engaged and passionate about the topic, and we learnt most about how others interact with the built environment from this approach. As others have reported, go alongs allow a rich experience, nuanced understanding and intersections between people and place to be explored (Gatrell 2013; Evans and Jones 2011). The approach also drew attention to the importance of the topic area and the need to continue to promote inclusive urban design that supports mobility for all, given the challenges presented by the built environment for many of our participants.

Practical Advice

Based on our experiences, this section provides an outline of some practical considerations when planning to undertake 'go-alongs' or walking interviews with older adults.

Participant and Route Selections

The selection of participants depends on the purpose of the 'go-along'. Here, we focus on walking as the means of the go along but there is potential for other forms of mobility within the 'walking environment' such as mobility scooters or wheelchairs, as well as other modes of transport such as bicycles, public or demand responsive transport or motor vehicles. The selection of participants and the mode of 'go-along' need to be appropriate for the research topic and design. The interviewer will need to take into account the physical function of the participants as there are potential health and safety risks associated with this method (e.g. a fall during the walk). Studies should specify how long participants must be able to walk for independently so that participants are aware before consenting what the expectations are, although of course flexibility can be exercised with each participant.

The number of participants involved in each interview should also be considered. Having more than one participant in an interview can minimise discomfort between participants and the researcher. Having more than one participant taking part in the research may mean that the participants themselves are able to take on a greater role in the research, particularly in terms of collecting data by operating camera, for example (Silveirinha De Oliveira 2011). On the other hand, from a practical perspective, recording the interview and absorbing information can be problematic with more than one participant. If the participants already know each other, because they are part of a walking group, for example, group interviews may work better than a group of strangers who may have different abilities and route preferences.

Route selection is a key consideration in a go-along interview (Evans and Jones 2011). Evans and Jones (2011) provide a typology of approaches from the route being determined by the interviewer to being determined by the interviewee and suggest that this depends on which of the two is more familiar with the environment. Our recommendation for walking with older adults is to encourage them to choose a route, but to discuss this before beginning the walk. In one of our studies, we asked participants to fill out a 'personal projects' form (Little 1983) beforehand indicating places they regularly visited and used that to discuss potential locations or routes for a walk. Participant selection is the preferred method for reasons related to ensuring comfort and familiarity in the environment and reducing the risks that participants might be exposed to by following unknown routes. However, as with participant selection, the route selection should also be informed by the research questions. If the researcher wishes to compare what different people think about the same route or local area, it may be more appropriate to have a predetermined route. In any case, preparing a route is advisable in case the participant prefers not to select a route.

Researcher safety should also be considered. In two of our case study examples, walking interviews were undertaken by a lone researcher. Where this occurs, lone working principles should be applied, such as notifying others of expected start and end times and location. Having a set route agreed prior to the interview provides safety protection for the researcher and participant. A risk assessment should identify the possible concerns. Safety equipment should be carried (e.g. first aid kit and fully charged mobile phone). Walks should ideally be taken in busy public areas, although it is recognised that some may be undertaken in more remote locations and therefore it is helpful if the above safety considerations are applied. We are aware of rare instances where the researcher has had to halt the walk because of fears over personal safety. This has ethical issues because saying 'I am not safe' to a participant projects the researchers perspective onto the participant, may make them feel more vulnerable and impact their behaviour in a particular environment. Participants should always be able to end the interview whenever they wish.

Methodological Considerations

Clear instructions need to be provided to the participant before embarking on the walking interview, for example, whether they are to lead the interview in terms of route, how long the interview can be expected to last and any research role they may take on, such as taking photographs during the walk. It is expected that this information and consent process takes place in a stationary environment prior to the go-along. This also provides an opportunity for collecting demographic data on the participants taking part in the study. During walk-along interviews, interviews are most likely to be unstructured or semi-structured. Preparing questions can be helpful to encourage discussions on the research topic and provide a useful tool if the discussion is not flowing easily. However, a benefit of conducting interviews in the environment is that it can be used in an elicitation process to prompt more discussion that may not occur in room based settings.

Part of the walking interview process can be to observe participants. However, taking field notes can be difficult to do while moving. Documenting the interview can be carried out in a number of ways, including the use of mobile technology. Interviews can be audio recorded using digital voice recorders and clip-on microphones to leave researchers and participants hands-free. This enables for further documentation in terms of taking photographs. The route and location of photographs can be recorded using a GPS. Most modern smartphones have this capability so expensive equipment is not needed. This allows interview transcripts to be matched up with geographic locations to geolocate and contextualise the conversation during analysis. Researchers should be aware that participants may not want to be recorded in this way, and therefore other ways of documenting the interview need to be considered. Older participants might not be used to technology and may not wish to use it during the interview. A follow-up interview may take place with the participant (e.g. at home after the walk) where additional activities can take place to further elicit, document and clarify points made during the walk. This may be particularly useful if the researcher is unable to audio record the interview or the mobility form does not allow for safe interviewing during the go along. For example, in two studies using bike-along interviews, the complete go along was videotaped and watched by the researcher and participants afterwards (Jones et al. 2016; Van Cauwenberg et al. 2018). A video-elicitation interview was then used to gather (additional) data.

Equipment and Recording the Interview

The equipment and technology used to record and document the interview will depend upon the 'go-along'. There are different considerations for interviews taking place while walking, compared to driving, for instance. It will also depend on what is suitable for the participant, and how confident they are using technology if they are to undertake a role in collecting information.

As previously mentioned, digital voice recorders with a clip-on microphone may be used to record the discussion between participants and researchers. Researchers should be mindful of weather conditions and provide wind guards for the microphones. Researchers should also be aware that the conversation may not be captured in its entirety, not only due to weather conditions, but also voices of passers-by and other sounds, such as traffic or construction works that are taking place in the environment. Using a notebook to document field notes may be useful in particularly noisy settings.

The development of mobile phone technology, i.e. smartphones, means there is now a plethora of applications to assist with the collection of data during a walk-along interview. Taking photographs using the smartphone camera can mean that photographs are automatically geo-tagged and can be readily plotted on maps. This enables easier documenting of the interview. It allows for interview questions and responses to be mapped using GPS and GIS. It is useful to visualise these data using mapping techniques.

Video cameras might also be used to record the interview. This covers both audio and visual documentation. As with smartphone technology, the development of cameras for action sports means that video cameras are small enough and wearable to use during a walking interview (e.g. GoPro). Such technologies present new opportunities for visual and audio analyses. More sophisticated specialist devices also allow potential for mixed methods approaches to understanding how people react to different environments through walking interviews alongside bio-sensing, for example (Osborne and Jones 2017).

Conclusions

We have outlined the potential of go-along or walking interviews as a geographical research method, detailed three example studies where we have used this method and provided some reflections and recommendations based on these experiences. The method is inherently suited to researching walking, particularly when the role of the built environment in influencing walking behaviour is of interest. Many studies seek to quantify aspects of the built environment which influence walking, but qualitative research *in place* provides a richer, more nuanced understanding of the intersection between people, behaviours and the built and social environments in which they live, which cannot be quantified or measured using methods more removed from the participant and environment. With increasing availability of crowd-sourced, scraped data from mobile devices through volunteered or non-volunteered geographic information (e.g. Google, Strava), it is important to remember the context in which this data is used and realise the implications of using such data without adding the perspectives of those whose mobility we seek to understand. We would not have been aware of many of the things we learnt about older adults' walking without using this method. That is not to say it is the only appropriate method, but that it can be important and complementary to other approaches. These conclusions and observations are based on our experiences. Many others will have different perspectives and experiences to share, but the hope is that by sharing these experiences others can learn and enjoy walking as a method.

Acknowledgements Angela Curl and Sara Tilley would like to acknowledge the support of the OPENspace Research Centre and the wider project teams involved in the fieldwork referred to in this chapter including Carol Maddock, Eva Silveirinha de Oliveira, Rick Houlihan and Chantelle Anandan for assistance with and reflections after interviews. They are incredibly grateful to the older people that gave their time and insights for the studies. The interviews for Case Studies 1 and 3 were undertaken at the OPENspace Research Centre, Edinburgh College of Art, University of Edinburgh. Case Study 1 was funded through the Engineering and Physical Sciences, Economic and Social, and Arts and Humanities Research Councils (grant reference number EP/K037404/1) under the Lifelong Health and Well-being Cross-Council Programme. Case Study 3 was part of the Getting Outdoors, Falls, Ageing and Resilience (GoFAR) project led by Marcus Ormerod at the University of Salford and was funded through the Medical Research Council (grant reference G1002782/1) as part of the Lifelong Health and Well-being (LLHW) Cross-Council Programme. The LLHW programme and funding partners had no role in the design, collection, analysis or interpretation of data; in the writing of the manuscript or in the decision to submit the manuscript for publication.

References

Butler, M., and S. Derrett. 2003. The Walking Interview: An Ethnographic Approach to Understanding Disability. *Internet Journal of Allied Health Sciences and Practice* 12 (3). Retrieved from http://nsuworks.nova.edu/ijahsp/vol12/iss3/6.

Carpiano, R.M. 2009. Come Take a Walk With Me: The "Go-Along" Interview as a Novel Method for Studying the Implications of Place for Health and Well-Being. *Health & Place* 15 (1): 263–272.

Centers for Disease Control and Prevention. 2013. U.S. Physical Activity Statistics.

Cook, S., A. Davidson, E. Stratford, J. Middleton, A. Plyushteva, H. Fitt, et al. 2016. Co-Producing Mobilities: Negotiating Geographical Knowledge in a Conference Session on the Move. *Journal of Geography in Higher Education* 40 (3): 340–374.

Cummins, S., S. Curtis, A.V. Diez-Roux, and S. Macintyre. 2007. Understanding and Representing "Place" in Health Research: A Relational Approach. *Social Science & Medicine* 65 (9): 1825–1838.

Curl, A., C.W. Thompson, P. Aspinall, and M. Ormerod. 2016. Developing an Audit Checklist to Assess Outdoor Falls Risk. *Proceedings of the Institution of Civil Engineers: Urban Design and Planning* 169 (3): 138–153.

De Vries, S. 2010. Nearby Nature and Human Health: Looking at Mechanisms and Their Implications. In *Innovative Approaches to Researching Landscape and Health: Open Space People Space 2*, ed. Catharine Ward Thompson, Peter Aspinall, and Simon Bell. Abingdon: Routledge.

European Commission. 2013. Special Eurobarometer 412 Sport and Physical Activity.

Evans, J., and P. Jones. 2011. The Walking Interview: Methodology, Mobility and Place. *Applied Geography* 31 (2): 849–858.

Gatrell, A. 2013. Therapeutic Mobilities: Walking and "Steps" to Wellbeing and Health. *Health & Place* 22: 98–106.

Ghekiere, A., J. Van Cauwenberg, B. de Geus, P. Clarys, G. Cardon, J. Salmon, et al. 2014. Critical Environmental Factors for Transportation Cycling in Children: A Qualitative Study Using Bike-Along Interviews. *PLoS ONE* 9 (9): e106696.

Goetz, A.R., T.M. Vowles, and S. Tierney. 2009. Bridging the Qualitative–Quantitative Divide in Transport Geography. *The Professional Geographer* 61 (3): 323–335.

Jones, P., and J. Evans. 2012. The Spatial Transcript: Analysing Mobilities Through Qualitative GIS. *Area* 44 (1): 92–99.

Jones, P., G. Bunce, J. Evans, H. Gibbs, and J. Hein. 2008. Exploring Space and Place with Walking Interviews. *Journal of Research Practice* 4 (2), Article D2.

Jones, P., R. Drury, and J. McBeath. 2011. Using GPS-Enabled Mobile Computing to Augment Qualitative Interviewing: Two Case Studies. *Field Methods* 23: 173–187.

Jones, T., K. Chatterjee, J. Spinney, E. Street, C. Van Reekum, B. Spencer, et al. 2016. *Cycle BOOM. Design for Lifelong Health and Wellbeing. Summary of Key Findings and Recommendations*. Retrieved from http://www.cycleboom.org/summary-report/.

Kusenbach, M. 2003. Street Phenomenology: The Go-Along as Ethnographic Research Tool. *Ethnography* 4 (3): 455–485.

Little, B.R. 1983. Personal Projects: "A Rationale and Method for Investigation." *Environment and Planning B: Planning and Design*, 15 (3): 273.

Merriman, P. 2014. Rethinking Mobile Methods. *Mobilities* 9 (2): 167–187.

Miaux, S., L. Drouin, P. Morency, S. Paquin, L. Gauvin, and C. Jacquemin. 2010. Making the Narrative Walk-in-Real-Time Methodology Relevant for Public Health Intervention: Towards an Integrative Approach. *Health & Place* 16: 1166–1173.

Michael, Y.L., E.M. Keast, H. Chaudhury, K. Day, A. Mahmood, and A.F.I. Sarte. 2009. Revising the Senior Walking Environmental Assessment Tool. *Preventive Medicine* 48 (3): 247–249.

Millington, C., C. Ward Thompson, D. Rowe, P. Aspinall, C. Fitzsimons, N. Nelson, and N. Mutrie. 2009. Development of the Scottish Walkability Assessment Tool (SWAT). *Health & Place* 15 (2): 474–81.

Morris, J.N., and A.E. Hardman. 1997. Walking to Health. *Sports Medicine* 23 (5): 306–332.

Nyman, S.R., C. Ballinger, J.E. Phillips, and R. Newton. 2013. Characteristics of Outdoor Falls Among Older People: A Qualitative Study. *BMC Geriatrics* 13 (1): 125.

Osborne, T., and P.I. Jones. 2017. Biosensing and Geography: A Mixed Methods Approach. *Applied Geography* 87: 160–169.
Phillips, J., N. Walford, A. Hockey, N. Foreman, and M. Lewis. 2013. Older People and Outdoor Environments: Pedestrian Anxieties and Barriers in the Use of Familiar and Unfamiliar Spaces. *Geoforum* 47: 113–124.
Sallis, J., R. Cervero, W. Ascher, K. Henderson, M. Kraft, and J. Kerr. 2006. An Ecological Approach to Creating Active Living Communities. *Annual Review of Public Health* 27: 297–322.
Scheffer A.C., M. Shuurmans, N. van Dijk, et al. 2008. Fear of Falling: Measurement Strategy, Prevalence, Risk Factors and Consequences Among Older Persons. *Age and Ageing* 37 (1): 19–24.
Silveirinha De Oliveira, E. 2011. *Immigrants and Public Open Spaces: Attitudes, Preferences and Uses*. PhD thesis, University of Edinburgh. Available at https://www.era.lib.ed.ac.uk/handle/1842/7953.
Vallée, J. 2017. The Daycourse of Place. *Social Science & Medicine*. In press. https://doi.org/10.1016/j.socscimed.2017.09.033.
Van Cauwenberg, J., V. Van Holle, D. Simons, R. Deridder, P. Clarys, L. Goubert, et al. 2012. Environmental Factors Influencing Older Adults' Walking for Transportation: A Study Using Walk-Along Interviews. *The International Journal of Behavioral Nutrition and Physical Activity* 9 (1): 85.
Van Cauwenberg, J., V. Van Holle, I. De Bourdeaudhuij, P. Clarys, J. Nasar, J. Salmon, et al. 2014a. Physical Environmental Factors that Invite Older Adults to Walk for Transportation. *Journal of Environmental Psychology* 38: 94–103.
Van Cauwenberg, J., V. Van Holle, I. De Bourdeaudhuij, P. Clarys, J. Nasar, J. Salmon, et al. 2014b. Using Manipulated Photographs to Identify Features of Streetscapes That May Encourage Older Adults to Walk for Transport. *PLoS ONE* 9 (11): e112107.
Van Cauwenberg, J., I. De Bourdeaudhuij, P. Clarys, J. Nasar, J. Salmon, L. Goubert, and B. Deforche. 2016. Street Characteristics Preferred for Transportation Walking Among Older Adults: A Choice-Based Conjoint Analysis with Manipulated Photographs. *International Journal of Behavioral Nutrition and Physical Activity* 13 (1): 6.

Van Cauwenberg, J., P. Clarys, I. De Bourdeaudhuij, A. Ghekiere, B. de Geus, N. Owen, and B. Deforche. 2018. Environmental Influences on Older Adults' Transportation Cycling Experiences: A Study Using Bike-Along Interviews. *Landscape and Urban Planning* 169: 37–46.
Wallenius, M. 1999. Personal Projects in Everyday Places: Perceived Supportiveness and the Environment and Psychological Well-Being. *Journal of Environmental Psychology* 19 (2): 131–143.

Part IV

Futures

9

Exploring How Older People Might Experience Future Transport Systems

Helen Fitt

Future Transport Systems

The chapters in this book very adeptly demonstrate that the transport system to which a person has access can make a substantial difference to their physical and mental health, their social opportunities, and their quality of life as they age. Research like that described in this book can help us to manage transport systems to meet the needs of older people and ageing populations. But our transport systems are changing. We are increasingly being described as on the cusp of some of the biggest transport disruptions our society has faced since the commercialisation of the motor car over a hundred years ago (Bansal et al. 2016; Graves 2017; RAC 2016). Autonomous (or driverless) vehicle technology is being rapidly developed, and visions of a primarily autonomous vehicle fleet are becoming increasingly credible. Autonomous vehicles are being

H. Fitt (✉)
Department of Geography, University of Canterbury,
Christchurch, New Zealand
e-mail: helen.fitt@canterbury.ac.nz

© The Author(s) 2018

A. Curl and C. Musselwhite (eds.), *Geographies of Transport and Ageing*,
https://doi.org/10.1007/978-3-319-76360-6_9

trialled in different conditions around the world, are being encouraged in different legal and political jurisdictions, and their commercialisation is being planned by a number of large industry players including Ford, GM, Renault-Nissan, Daimler, Volkswagen, BMW, Waymo, and Toyota (Bloomberg Philanthropies 2017; Muoio 2017). Many of these companies are also developing their electric vehicle offerings and, in 2017, Volvo announced an end to its sales of combustion-only cars (Vaughan 2017). In addition, several countries (including France, the UK, Norway, and China) have set in place targets or strategies to eliminate internal combustion-only cars from their national fleets (Petroff 2017; *The Economist* 2017). These changes to mobility sit alongside ongoing changes to the ways we interact in person and remotely, especially through increasing sophistication of online technologies. In this context, it is appropriate to start considering what our transport future might look like, how future possibilities might affect older people and ageing populations, and what we can do to ensure that new technologies are adopted in socially beneficial ways.

In 1932, English writer HG Wells lamented the lack of foresight that had gone into planning for the implications of car travel. He said:

> See how unprepared our world was for the motorcar. The motorcar ought to have been anticipated at the beginning of this century; it was bound to come. It was bound to be cheapened and made abundant. It was bound to change our roads, take passenger and goods traffic from the rails, alter the distribution of our population, congest our towns with traffic. … Did we do anything to work out any of these consequences of the motor car before they came? Not much. We did nothing to our roads until they were choked. We did nothing to adjust our railroads to fit in with this new element in life, until they were overtaken and contemplating the possibilities of bankruptcy. … In the case of the motor car, we have let consequence after consequence take us by surprise, then we have tried our remedies, belatedly; and exactly the same thing is happening in regard to every other improvement in locomotion and communication. (Wells 1932)

If we do not consider the implications of future changes to our transport systems, we risk perpetuating the reactionary stance HG Wells lambasted. We can be confident, for example, that automation of the vehicle fleet would have both direct and indirect implications for society. Direct implications would affect transport infrastructure (including road markings and traffic signals), congestion, urban form and population distribution, modal share, and wider social dynamics. Indirect implications would be manifested through the recursive interlinkages between transport systems and society. For example, changes to urban form and population distribution could trigger subsequent changes to transportation systems and social dynamics. The exact qualities of direct and indirect implications of automation are, as yet, uncertain. Many published opinions seem to be somewhat polarised, and both utopian and dystopian visions, each with their own implications for an ageing population, are currently common (Alessandrini et al. 2015; Edwards 2012; Gleave 2017; Haratsis 2016; Margolis 2017; Plautz 2016; Polonetsky and Claypool 2016).

Utopian visions commonly include a range of potential benefits of autonomous vehicles. First, autonomous vehicles may substantially reduce road accident rates and facilitate full participation in social life for those who are currently unable to drive (including many older people) (Alessandrini et al. 2015; Anderson et al. 2014a; Cavoli et al. 2017; Harper et al. 2016; Krueger et al. 2016; Shergold et al. 2016). Vehicle sharing may become the norm thus substantially reducing the number of cars needed in the vehicle fleet and the cost of accessing private motorised transport (Bansal et al. 2016; Cavoli et al. 2017; Clark et al. 2016; Litman 2017). The space requirements of automotive traffic in terms of both road space and parking may decline considerably, freeing up more valuable urban land for other uses, including affordable housing and urban green spaces (Anderson et al. 2014b; Cavoli et al. 2017; Graves 2017; Margolis 2017; Wakabayashi 2017). Admittedly, the quantum of travel might increase, but traffic systems could be more efficient and electric vehicles can have very low emissions, so the current problems created by mass automotive travel could be largely eradicated (Alessandrini et al. 2015; Anderson et al. 2014b; Litman 2017).

In contrast, dystopian visions paint a picture of massively increased traffic volumes (Anderson et al. 2014a; Graves 2017; Litman 2017). When a vehicle drives itself, it is argued, its passengers may find travel time less of an imposition than they do when they have to concentrate on driving (Bansal et al. 2016; Cavoli et al. 2017; Edwards 2012; Litman 2017). As a result, people might move further and further from urban centres causing sprawl to accelerate (Anderson et al. 2014a; Bansal et al. 2016; Cavoli et al. 2017; Edwards 2012). An increase in the number of people able to travel independently (including children and older adults with travel restrictive medical conditions) would result in further increases in traffic volumes (Harper et al. 2016; Litman 2017). Private vehicle ownership models might persist, in which case, the vehicle fleet would grow (KPMG 2013). Expensive vehicles (and regular costly software updates) may remain unaffordable for some members of society, and late adopters of new technology [which might include some but not all older people (Mansvelt 2017)] may struggle to keep up with new ways of getting around. Those excluded from new transport systems may find themselves unable to access essential goods and services in a decentralised, sprawling, and traffic-dominated urban environment (Blumgart 2013; Cavoli et al. 2017; Litman 2017). Hackers, criminals, and terrorists could target both individual vehicles and wider traffic systems through introducing viruses or rogue communications into highly connected networks (Anderson et al. 2014a; Clark et al. 2016; Department for Transport 2015; Duncan et al. 2015; Kyriakidis et al. 2015; Litman 2017; Margolis 2017).

Both the utopian and dystopian discourses have spatial and social implications, thus making them highly relevant to a consideration of the geographies of transport and ageing. Both of these discourses may also, at first glance, seem credible but what they fail to account for is the nuance, complexity, and multiplicity of social life. In recent decades, social scientists have increasingly rejected notions of homogeneity and stasis in social life (Adey 2010; Bennett 1999; Butler 1999; Clarke et al. 2003; Gamson 2003; Halnon and Cohen 2006; Law 1999; Maffesoli 1996; Marshall and Rossman 2011; Middleton 2009; Muggleton 2007; Pile 2010). Work on subcultures, postmodernism, queer theories, feminism, poststructuralism, and mobilities has increasingly emphasised that social life is complex and variable and that singular narratives of social reality rarely represent lived experience (Adey 2010; Bennett 1999; Butler 1999; Clarke et al. 2003; Cloudsley 2007; Gamson 2003; Guell et al. 2012; Halnon and Cohen 2006; Law 1999; Maffesoli 1996; Shove et al. 2012). How then might we paint more nuanced and complex pictures of what future transport systems might be like and how they might integrate with the multiple, variable, and interwoven contexts of real lives?

Research that goes beyond utopian and dystopian binaries of the likely implications of vehicle automation is proliferating. However, social scientists (including human geographers) sometimes find research focusing on an uncertain future to be a profoundly slippery thing. The more we accept complexity and nuance, the harder it becomes to extrapolate a situation forwards to inform proactive decision making. This chapter contains a set of narratives that focus on future possibilities and are intended to prompt the kinds of thoughts and discussions that allow us to incorporate complexity and nuance in our considerations of the future.

Methods

In creating the narratives you are about to read, I have borrowed from three different social science approaches: foresight methodologies, mobilities scholarship, and social practice theories. First, foresight methodologies can help us to explore plausible metanarratives of transport. This chapter uses prospective scenarios devised by the New Zealand Ministry of Transport (MOT), using foresight methodologies, as a basis for further explorations. In the Future Demand foresight exercise, MOT considered the potential implications of two possible dynamics: differing social preferences for virtual or face-to-face connections, and changing energy costs relative to incomes and living expenses (Lyons et al. 2014).[1] The resulting 'double uncertainty matrix' (Shergold et al. 2015, p. 88) shows four plausible scenarios for life in 2042 and is illustrated in Fig. 9.1. These four scenarios form a foundation for the rest of this chapter.

Scenario and visioning exercises have recently been criticised for often conceptualising people as homogenous and rational actors (Bergman et al. 2017; see also Shergold et al. 2015). Although the MOT scenarios avoid some of the generalisations made in other exercises (Bergman et al. 2017), they still focus on broad national implications

[1]MOT has also conducted further foresight exercises (see, e.g., Ministry of Transport 2016); in the interests of brevity, this chapter focuses on the first exercise only.

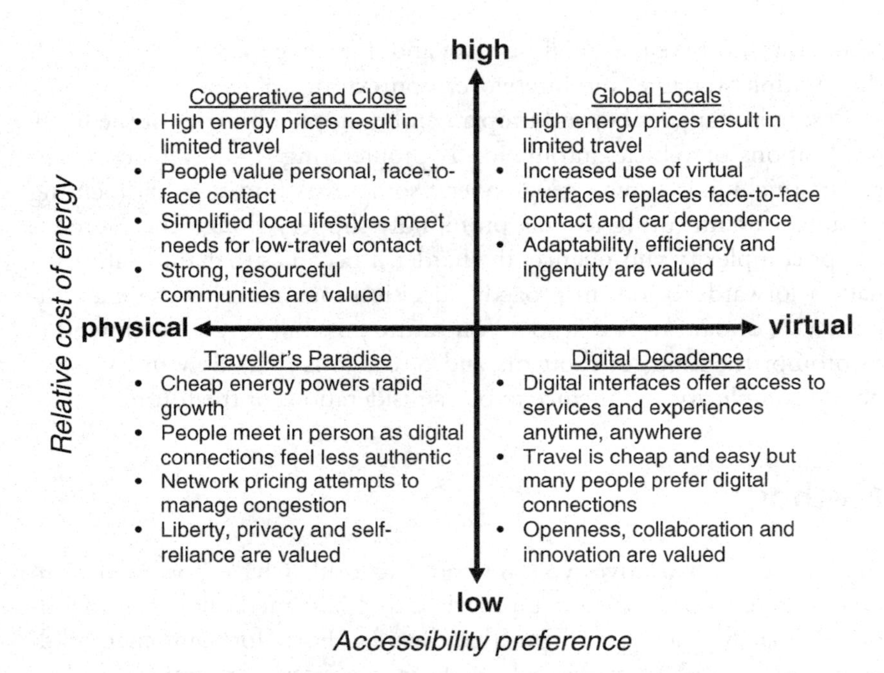

Fig. 9.1 Future demand double uncertainty matrix. Adapted from Lyons et al. (2014, p. 18) with permission from the Ministry of Transport

of changes in transport systems. To move beyond these, towards more socially and culturally nuanced understandings of implications for heterogeneous and complex people, I have leant on insights from social practice theories and mobilities scholarship. Social practice theories posit that transport practices should be understood as part of an evolving and holistic system in which there are elements of stability and change (Giddens 1979; Schwanen and Lucas 2011; Shove et al. 2012). Through considering which elements of systems might change, which might stay the same, and how different elements relate to one another, we can take a systematic approach to moving beyond metanarratives of transport. Leaning on mobilities scholarship complements this approach through considering practices to be multiply determined, variable, and contextual, as well as having embodied or phenomenological importance (Adey 2010; Büscher et al. 2011). The combination of scenarios, with insights from social practice theories and mobilities scholarship, can inspire the generation of plausible micronarratives of the lives of individual older people.

The stories that form the main body of this chapter focus on the everyday lives of 12 characters between the ages of 51 and 84. These characters are based on real research participants who took part in focus groups, detailed qualitative travel diaries, and individual interviews in 2013 and 2014. The original research in which participants took part was an in-depth qualitative investigation of the influences of social meanings on everyday transport practices. As such, it involved detailed explorations of people's daily practices and of the cultural, instrumental, affective, and experiential influences on those practices. A full explanation of the previous research, including literature review, methods, and findings can be found in Fitt (2016). This chapter uses the very detailed data collected in that original exercise to consider what the everyday lives of participants might be like in 2042 if they were living in a New Zealand that resembled one of MOT's four scenarios. Each participant's age has been increased by around 30 years, and elements of their personal attitudes and situations have been combined with the scenarios to generate the stories below. Basing the stories on real people makes it possible to show a wide (but realistic) variety of different circumstances, perspectives, and preferences with a minimum of speculation. These micronarratives encourage us to think through some of the more fleeting, nuanced, multiple, contradictory, and chaotic repercussions of broader changes in the everyday lives of real people.

It is important to note that the characters in the stories below are *based* on real people but are *not* real people. Some of their apparently enduring perspectives are real (and sometimes real quotes are used, indicated by double quote marks) but elements of their situations, characteristics, practices, and perspectives have been altered to create short, future-focused narratives. There are strong connections between the narratives and the real people on which they are based, but the narratives have been devised with the specific intention of demonstrating (without attempting to predict) the wide range of multiple, variable, and nuanced everyday lives that we could expect to encounter in MOTs future scenarios.

The Ministry of Transport's original scenarios do not have an explicit focus on age. New Zealand, along with many other countries, has an ageing population (Statistics New Zealand 2016; United Nations 2015). Specifically, exploring what the scenarios might mean for older people and for ageing societies, then, provides a relevant extension to the original work. This chapter starts to bring an age perspective to the scenarios, but the treatment here is necessarily partial. In the interests of brevity, I have only provided three character stories for each scenario. These stories are varied and do reflect some of the nuance and complexity of everyday life, but it would be easily possible to provide many more stories, each of which would be different. To expand the range of stories provided, imagine yourself, or your parents, or your children, partner, best friend, or colleagues in any or all of the scenarios and see how the resulting perspectives differ.

One important consideration in creating the stories here has been sensitivity to the feelings of the original participants on whom the characters are based. None of the story characters is mourning a spouse, none of them is experiencing major age-related decline (and those with some experiences of decline all have social support), and none of them is living in a nursing home, hospice, or palliative care facility. There is a rich literature exploring all of these topics, and further explorations of ageing in different scenarios and conditions could provide a rich resource for a society wanting to proactively manage its future possibilities.

Stories

Travellers' Paradise

In Ryan (82), Lily (61), and Ashley's (53) New Zealand, face-to-face interaction is the norm. The social media bubble has well and truly burst, and people choose to keep their personal data private and their social lives offline. People living in this New Zealand have access to cheap fuel, enjoy low living costs, and experience prolific congestion. The stories below demonstrate some of the range of different experiences that people might have in a New Zealand like this.

Lily (61) lives on the outskirts of the city where house prices are lower than in the centre. She used to own an old car for commuting to work, but she wrote it off in an accident last year and the insurance company wouldn't replace it. Of course, if she'd had fully comprehensive insurance, or a car with collision avoidance technology, then she wouldn't be carless now. But—like many people with low incomes—insurance and the latest technology are out of her reach. Since the accident, Lily has had problems with headaches and poor concentration, but she hasn't seen a doctor because doctors are expensive and would probably recommend time off work, which isn't really an option.

Lily's stoic about her situation, it is what it is, and it could be worse. After the accident, Lily moved house so that she could get a train to work; the train is much quicker than the bus and avoids a lot of the congestion on the roads. She liked her old neighbourhood better (it had more local facilities, more street life, and felt a bit less like a dormitory suburb) but while she's still working, the train trumps the appeal of shops, cafés, and a park. When she retires, Lily thinks, she'll move back to a bus suburb because she won't have to travel in peak hours and she won't have to go into the centre of town anywhere near as much.

Ashley (53) isn't thinking about where she should live when she retires, she's too busy planning a holiday. She's working from home today (which most of her colleagues do a couple of times a week to avoid the congestion and the peak hour road charges between the inner suburbs and the city centre). It's a lazy start, but if she went to work it would be at least another hour before she made it there so she's not feeling guilty about her lack of productivity (yet).

Ashley mostly uses public transport or pay-as-you-go car share cars depending on her destination and mood. Cars are more expensive than public transport, but they're usually quicker, unless your destination is on a direct train route. Besides, Ashley enjoys driving. She also acknowledges that there is a little bit of stigma involved in taking buses, because it is mostly poorer people who do so. She says that doesn't stop her from travelling by bus, but it's not her favourite way to travel. Ashley usually chooses electric cars when she can, because she knows it's better for the environment and for noise pollution (especially with some of those monster traffic jams).

Ryan (82) owns two different vehicles and uses whichever is most appropriate for the journey he's making. If he needs to pick up supplies to fix a minor problem at one of the rental properties he owns, he'll use the ute because it's got a decent amount of space in the back and it doesn't matter if it gets dirty and a little bit battered. On the other hand, if Ryan's going out socially or driving his wife to one of the more exclusive shopping areas that she likes to visit, then he takes the Mercedes because, while he doesn't really think of himself as refined, he likes to play that role sometimes. Ryan has always done a lot of driving, including some in quite challenging conditions, and he considers himself an excellent driver. He recently had a close call on a busy roundabout (but the collision avoidance technology in the Merc kicked in just in time). His wife is hassling him to get his eyesight tested again, but Ryan's pretty sure it was just a momentary lapse of concentration; no one's perfect no matter how skilled they are.

Global Locals

In a stark contrast to Ryan, Lily, and Ashley's New Zealand, Theo (66), Paul (81), and Daniel (84) all live in a New Zealand in which most people live their lives primarily online and transport is not very important for managing life's necessities.

Theo (66) is looking forward to retiring soon and spending more time with his daughter, Hannah. Hannah lives in Beijing, and the time difference means quality time together has been rare. But Theo's intending to celebrate his retirement by playing virtual chess with Hannah whatever time of day she's free. Given the quality of virtual interfaces these days, chatting online is really not that different to actually being in the same room as someone. Theo has spent most of his working life in telecommunications and has always been that person who friends and family ask for help when they can't make their phone or computer work. He's got his interface set up just how he likes it and has no qualms at all about fiddling with the settings to get the best experience.

Theo's also looking forward to getting outside more once he's retired. He's been working from home a lot for the last few years, and there's really no *need* to actually leave the house. He usually gets out for a bike ride or walk at the weekends, and it's great that there's so little traffic these days, but Theo misses the bush. Getting into the back blocks by public transport isn't really an option for weekends; once he's retired though, Theo's keen to go and explore all the lakes, rivers, and bush he hasn't visited for years. As he talks about this, his face is consumed by an ear-to-ear grin.

Paul (81) isn't grinning as he reflects on his retirement. He lives in a suburban unit that he says is about the size of a matchbox but still hard to heat in winter. He would go out, but there really isn't anywhere to go. The library, which was once a warm haven for book lovers and full of other living souls to be around, is now just an interface of clickable titles. Paul used to spend his weekends wandering around the shops, but there aren't any shops nearby anymore, and besides he doesn't feel safe going out anymore. After the first time he was mugged, Paul used to tell himself "I'm not giving into it. Because the moment you do that, they've won", but now he admits that they *have* won.

Last month, Paul signed up for a free online philosophy discussion group, and he's met a couple of people he gets on with quite well. Interacting online is new for him, but a couple of the members have decided to have online drinks on Friday evening and Paul has had some beer delivered in anticipation. He just hopes he'll be able to get his microphone working by then.

Daniel (84) knows he's at the privileged end of society. He has managed his money prudently, but he does value being able to sometimes make choices that don't make financial sense. He knows, for example, that the autonomous car share scheme he has joined is the most expensive one available and he's fine with that. The cars are quick to arrive when summoned, they're always immaculate, they have the best in-vehicle entertainment systems available, and they do make a bit of a statement. If you're going to go somewhere, Daniel thinks, you might as well enjoy the journey.

The one thing that concerns Daniel about the car share scheme is safety. Daniel has long been aware that society is not particularly forgiving of "rich tossers" in expensive cars. The car itself is pretty safe, and very hard to get into if you're not a registered scheme member, but Daniel knows it really wouldn't be hard for someone to follow him home and work out where he and his wife live. He's set his personal preferences for the vehicle scheme to their most courteous setting (so his ride won't cut people up or sneak into a park in front of someone else); there's no need to antagonise people…especially when you're enjoying the journey anyway.

Digital Decadence

Ollie (78), Tamara (63), and Quentin (83) live in a much more equal New Zealand. The government has fought to ensure that everyone has access to the goods and services they need and has declared that digital connectivity is a human right. Most cars are electric and autonomous, but some people still prefer traditional petrol vehicles that you actually have to drive yourself.

Ollie (78) has been a car enthusiast for many years but for a long time he didn't have the money to follow his passion. Now autonomous car sharing has become most people's preferred way to travel, older cars are relatively cheap. Ollie can't decide which of his cars is his favourite; it's probably between the BMW525i, the Mitsubishi Lancer with the additional turbo, and the Chevy Camaro IROC-Z. People have tried to talk him into converting his cars to electric motors but Ollie is having none of it. He is not a "tree hugger" and a car is supposed to go vvvvvvrrrrrrrmmmm.

Ollie particularly loves going on long road trips. He says "I just find when I go on a long trip…my mind kind of clears…you know, I'm in control to some extent. …You get into this kind of rhythm, you've got the natural…rubber based, iambic pentameter of the vvvvvvvvv [makes a car noise] …and I seem to kind of, get in a zone, I suppose". While Ollie is happy to comment on the ineptitude of octogenarians on the road, he cannot (or will not) contemplate a day when he might be too old to safely drive.

Tamara (63) has never understood the need for "huge chrome-wheeled, disgusting, big, fuel guzzling [vehicles]" and now that public transport is cheap, reliable, and door-to-door, she really doesn't get why anyone would bother with owning an old car. That said, there is a little part of her that misses the status of being able to drive. Other people used to see her as a bit of a girly-girl and doubted her technical savvy. Although she was never much into car maintenance, she felt proud to disrupt gender stereotypes by driving a manual car, not an automatic. Of course, most people don't drive at all now so that's not really a distinction that she can still claim, which she thinks is a bit of a shame.

Tamara admits, though, that some technological changes have made it easier to be a single, older female. She has recently taken several digital immersion trips to far-off places, and she particularly enjoys virtual trips to places where she might feel a bit unsafe if she went in person (not everywhere is as safe as Tamara's New Zealand). Tamara is currently contemplating becoming a virtual volunteer for a medical charity operating in live conflict zones when she retires. She's got a few years of work to go before she can do that, but she's got relevant skills, would like to help and maybe (although she doesn't really admit it) there's a part of her that thinks that would really show the boys who's tough.

Virtual health care is an opportunity for Tamara, but Quentin (83) remains unconvinced about the digital clinic he attends. He's been struggling with his memory recently, and his doctor has told him he has a mild, age-related, cognitive impairment. Quentin would rather summon a car and go talk to the doctor face-to-face, but that's not an option because he's been assigned to a clinic at the other end of the country. It's the best care for his condition, but that doesn't mean he likes the method of delivery.

The proliferation of online services has meant that Quentin and his wife, Hazel, have been able to move out of the city. Things like shopping, banking, and voting can be done from pretty much anywhere although Hazel has to keep an eye on what Quentin's doing online to make sure he doesn't order the food twice or forget to pay a bill. Quentin and Hazel go for long walks most days and enjoy regular visits from their great-grandchildren (who sometimes come in the family's driverless car without their parents). Quentin still prefers to socialise face-to-face, and he's delighted that a nearby village has a 'retro-café' where people go to chat. For Quentin, walking to the café is a matter of principle, even though it takes him about an hour and a half each way.

He says that activity is important for maintaining his health, and just because he could get an autonomous pod, doesn't mean he should.

Cooperative and Close

In Reece (72), Zoe (79), and Kelly's (51) New Zealand, simple tastes and low technology use are more the norm. In response to high energy costs and decreasing trust in digital media, this New Zealand has embraced a simple, resourceful, local lifestyle in which travel is kept to a minimum and people live and socialise primarily within strong local communities.

Zoe (79) is a fan of the simple lifestyle. When she was working, life seemed to get constantly faster; now she likes what she calls "the slower pace of life" and rarely misses any of the gadgets she used to use. Zoe lives most of her life within a couple of kilometres of her home and knows most of her neighbours and the owners of most of the local businesses. Community members often share things like garden tools and seeds. Zoe had a fall recently, which made her pretty nervous about walking to the shops, but one of her neighbours, Kim, has been an absolute star. When Kim heard that Zoe had had a fall, she suggested that they could do their local errands together. Zoe knows that Kim has struggled with mental health issues in the past and is glad that Kim probably values the company as much as she does.

When Zoe and Kim walk to the shops, they pass a gated community with private security guards who open and close the gate for shuttle buses and the occasional private car. The people who live behind the gates aren't part of Zoe and Kim's community, and Zoe thinks they're "arrogant" and "up themselves". When she walks past the gate, if there's a vehicle going in or out, she pretends to be more frail and slow than she really is because the residents of the gated community make no contribution to society and she secretly enjoys making them wait.

Reece (72) and his wife, Meredith, moved into the city centre just before Reece's 70th birthday. They were doing fine in their old suburban neighbourhood, but the allure of the city was just too much. Being able to go out beyond their own neighbourhood in the evenings without having to get into a crowded minibus seemed enough to justify the higher house prices in the centre. Walking and biking are both options when everything is so close, and Reece and Meredith say it's nice to be able to go out in the evening without having to think about how they're going to get home. Meredith recently bought an electric bike, and she thinks it's the best thing ever. She has the freedom to go wherever she likes without getting sweaty or tired, and it stops her from getting left far behind when she and Reece bike somewhere together. Reece has no intention of getting an electric bike for himself; he's proud of his fitness and doesn't want to stop biking properly and risk putting on weight. Besides, he thinks electric bikes are cheating.

Kelly's (51) still many years from retirement and loves the very specialist (although not particularly well-paid) design job that she does. The only downside to her chosen career is that she can't pursue it within her own neighbourhood. Getting to work by bus is ridiculously expensive and she's often late because the service is unreliable. She does her best to get some work done on the bus but it's difficult, especially on the days when she can't get a seat. One of her colleagues recently bought a moped, and Kelly has always thought that it would be "quite cool" being able to "zip around", especially on warm, dry days. Living out where Kelly does though, it's just too dangerous on the roads. Kelly is emphatic that she doesn't want to die.

Kelly's long commute means she doesn't usually have time for much more than dinner and bed in the evenings. This means she isn't really part of her community, and she knows her neighbours think that's a little strange. When she does retire, she'll have to try to integrate herself into a community somehow if she wants to have any ongoing social connections. Working in the city does have its advantages though; the boutique shops near Kelly's work have much better fresh foods than the shops near her home. Kelly would never travel just to buy those things, but when she's so nearby for work it seems a no-brainer.

Discussion

This chapter is largely composed of stories that combine some of the characteristics and preferences of real research participants with some of the main elements of the New Zealand Ministry of Transport's future scenarios. The stories are intended to prompt thought and discussion on what feasibly could happen in future and on what is desirable. Through combining metanarratives and micronarratives, the chapter allows us to consider potential futures from a range of different perspectives. For example, it allows us to consider how changes in major structural elements of a society (like transport systems, urban form, and approaches to wealth inequality) influence the practices and experiences of its older members. At the same time, it allows us to reflect on some of the agency, idiosyncrasies, and individualities that also influence everyday practices and experiences.

The stories above have been written in a way that incorporates some important issues associated with older people, mobility, and transport. Many of these issues are described in more detail in extant literature, but it is appropriate here to highlight some of these themes and their relevance for our consideration of future transport systems. As social science increasingly learns to consider future possibilities, we will also need to learn to go even further beyond what we already know, identifying and moving past some of the assumptions that we commonly make about the social, spatial, and temporal configurations of daily life (Bergman et al. 2017).

Age is clearly an important element of many of the stories above. Each of the characters is based on a real research participant and is described as being the (approximate) age that that participant will be in 2042. There are two dynamics associated with character age (and ageing) that have particularly informed the development of the stories. First, previous research has demonstrated that the rhythms of daily life are intricately connected to age and to life stage (Chatterjee et al. 2013; Harada and Waitt 2013; Krueger et al. 2016). Seven of our twelve characters are retired in 2042, five are working. People who work often face tight constraints on their everyday mobility, particularly with regard to working hours and locations. In the stories above, Lily and Kelly particularly struggle with their quite rigid commutes, while Theo and Ashley are able to work from home and so appear to have a little more freedom. Theo's story, however, does allude to a level of dissatisfying immobility and confinement relating to rarely leaving the house and not having regular access to the outdoors. In contrast, our retired characters rarely mention constraints around the timing or destinations of their movements. Ollie loves road trips, Quentin and Hazel enjoy long walks, and Lily (who is not yet retired) looks forward to a time when some of her constraints will be gone and she can move to a place she would prefer to live. Of course, retired people may also face time and location constraints on their movement (possibly relating to ongoing non-work commitments or concessionary fare limitations on public transport services) but retirement is often a point at which mobility constraints change. When we think about ageing, then it is relevant to think about both pre- and post-retirement patterns of mobility and how they might change.

The second major dynamic associated with age in the stories above is that of physical and cognitive decline. Such decline is commonly experienced by ageing individuals; for example, a third of people aged over 65 may end their lives with dementia (Lakey 2010) and a similar proportion may experience one or more falls (Hawley-Hague et al. 2014). Several of the characters above are dealing with age-related decline and the implications that has for their mobility. For example, following a fall, Zoe is nervous about walking in her local area, Quentin and his wife are just starting to deal with the effects of a cognitive impairment, and it's really not clear whether Ryan's near miss while driving around a roundabout was, as he says, 'just a momentary lapse of concentration' or something more significant. These characters live their lives at quite different geographical scales according to the wider context of each New Zealand. Zoe lives in a local, community-oriented setting where essential services are primarily within walking distance; Quentin lives in a rural environment with goods and services delivered chiefly through virtual connections; and Ryan lives in a more urban environment where he relies substantially on his car. In the stories presented, each character who is experiencing age-related decline has some social support; Quentin and Ryan have their wives, and Zoe has her neighbour, Kim. This may not always be the case, and it would be appropriate to revisit each scenario considering the implications of different kinds of age-related decline and different social support contexts.

Differences in transport systems (and how people use them) are a another major theme informing the stories above, and here in particular, the differences incorporate both broad social currents and some individual preferences and idiosyncrasies. Some of the broad currents include the kinds of transport that are available (and accessible to different individuals) and how important physical mobility is to full participation in social life. Unlike ageing, these themes are a significant focus of the original Ministry of Transport scenarios, with each scenario having certain transport characteristics. For example, in the Travellers' Paradise scenario, physical mobility is important and the transport system features cheap fuel and prolific congestion alongside mass transit rail services and rapid bus networks. In contrast, physical mobility is much less important and virtual mobility more important in the Global Locals scenario.

The characters in the stories have access to different transport options depending on the overarching scenario under which they live and on their personal circumstances. Often characters can choose between available (although usually limited) options depending on their own preferences. Some characters drive while others use autonomous vehicles, some walk, bike, or use public transport; Tamara even travels virtually to far-off places; and Paul rarely braves going out. Although most of the participants are described in terms of a single, primary mode of transport, I have tried to make it clear through the stories that the complexities of life often mean that people use different modes in different circumstances (see also Shaw and Docherty 2014). For example, Ashley chooses between public transport (trains and buses) and driving. Some of those who use motor vehicles use combustion-driven vehicles and some use electric; some own their own cars, while others are members of car sharing schemes. Ryan, in the Travellers' Paradise scenario, even owns two different vehicles that he uses for different kinds of trips.

Beyond main themes related to age and transport, these stories also highlight a number of other dynamics worthy of further discussion. For example, wealth, and particularly the ways in which wealth inequalities are managed are important features of many of the scenarios. The Digital Decadence scenario is the only one that incorporates explicit policy goals relating to social equity. In the other scenarios, we see clear differences in the choices available to those in upper and lower wealth brackets. Perhaps most striking is the difference between Lily and Ryan's transport choices in the Travellers' Paradise scenario. Lily was involved in a road traffic accident that might have been prevented if she had been able to afford collision avoidance technology. Her car was written off and, because she had not been able to pay for fully comprehensive insurance, she could not afford to replace it. She was suffering ongoing health issues as a result of the accident, could not afford to see a doctor or take time off work, and had had to move out of her preferred neighbourhood to facilitate commuting by public transport. In contrast, Ryan had a near miss while driving but his car's collision avoidance technology enabled him to avoid an accident and, other than his wife urging him to get his eyesight tested, little in his life changed. Very similar initial events resulted in quite different outcomes for Ryan and Lily, primarily because of their different access to financial resources.

Different urban forms, and differential access to goods and services, are also key features of the stories above. Combustion-driven vehicles have historically been associated with widespread changes to urban form, and particularly to decreasing urban density and increasing sprawl (Blumgart 2013; Edwards 2012; Rao et al. 2007). Both the adoption of autonomous vehicles and increases in remote access to goods and services could contribute to further low-density expansion away from urban centres. In the Digital Decadence scenario, for example, online access to services such as food deliveries, banking, voting, and medical care allowed Quentin and his wife to move out of the city. Such changes may be liberating for individuals like Quentin, but negative repercussions might include the disappearance or dispersal of physical places like doctors' surgeries, libraries, and shops. If such places become harder to access, people like Paul, in the Global Locals scenario, may lose opportunities for social connection and become increasingly isolated. Changes in the accessibility of physical places will be particularly concerning if some groups, like those with cognitive impairments, do not have access to the autonomous vehicles and online services that allow others to live comfortable lives more remotely. The ability to age in place—that is, to grow old without having to move out of a familiar house, neighbourhood, and community—has become increasingly regarded as important to healthy ageing. Urban form that supports independent living by people with a range of physical and cognitive capabilities is important to successful ageing in place (Burton et al. 2011). The stories above contain numerous hints about how changes in transport systems could influence people's abilities to age in place, including through their influences on access to goods and services, social opportunities, quality housing choices, streetscapes, crime, community segregation, local economies, and more.

There are many more dynamics in the stories above that we could discuss in a lot more detail. We could, for example, explore issues associated with technological competence. Different levels of comfort, experience, and expertise with technology may influence the opportunities that individuals are able to access. For example, in the stories above, Theo is a technology expert and he takes full advantage of the gadgets available to him to facilitate social connections (such as playing chess with his daughter in Beijing). Paul, on the other hand, has the technology he needs to connect with other people, but is not really sure how to use it or even what to expect if it works. There is a common assumption, not always supported by research, that older people struggle more with technology than younger people (Demiris et al. 2004; Lyons et al. 2017; McMahon et al. 2016; Mihailidis et al. 2008; Shergold et al. 2016). As with other characteristics discussed in this chapter, we can expect older people to have a complex range of different engagements with technology (Mansvelt 2017). However, as transport and related technologies change, the technological competences that people need to possess will also change. For example, what competences would we need to manage the software updates that might be required for autonomous vehicles to continue working? If your computer has ever launched a major update while you were in the middle of delivering a PowerPoint presentation, or your phone has claimed to be too old to launch a certain app, or you have ever wondered whether a suggested update, app, or piece of software is legitimate, safe, and necessary, or is riddled with malware and hidden costs, think about how a similar situation could play out in a transport context. How would society (including its older members) manage regular, perhaps confusing, and potentially costly software updates? Could this prevent some people from managing autonomous vehicle ownership? Similarly, can we be sure that all members of our society would have the necessary skills to use mobility-as-a-service platforms?

Further, while some of the characters in the stories are members of car sharing schemes, we have not considered in this chapter what makes car sharing schemes work, where the cars are located, how they are booked, accessed, and maintained, and how they may or may not contribute to equitable accessibility (see, e.g., Dowling and Kent 2015). An autonomous vehicle future, it is commonly suggested, will facilitate changes in models of ownership, sharing, and usership (Bansal et al. 2016; Cavoli et al. 2017; Clark et al. 2016; Litman 2017); more research exploring this topic could help inform proactive management of future transport possibilities. We could also go on to explore issues associated with charging electric vehicles; how fast or slow it is, how expensive it is, or how well it fits with the routines of everyday life for different characters. Finally, there is no discussion in this chapter of how autonomous vehicles fit in and interact with other modes of transport. As with ageing though, there are rich bodies of literature that cover most of these topics and these could fruitfully be combined with both micro- and macronarratives for a richer understanding of future possibilities.

Scenarios such as those produced by the New Zealand Ministry of Transport can provide excellent starting points for considering future possibilities. This chapter, though, tries to add depth, nuance, and complexity to those scenarios by combining them with data about real people to provide rich and thought-provoking stories. People are influenced by the wider social and technological metanarratives that shape the worlds in which they live, but people also have their own idiosyncrasies, preferences, and particularities. Quentin chooses to walk for an hour and a half each way to a café, Tamara chooses to work as a virtual volunteer for a medical charity, and Zoe chooses to inconvenience those she perceives as arrogant and asocial residents of a gated community. Different people might respond to the same circumstances in very different ways, and when we incorporate some of these idiosyncrasies into our discussions about the future, our proactive management of future possibilities is likely to be enhanced.

Ultimately, the complex and recursive links between possible transport futures and people's possibilities, practices, and preferences offer rich potential for much wider discussions than could ever be included in a single book chapter. This chapter, however, offers a starting point for thinking about where we might go from here, both in terms of methodological strategies for thinking about the future and in terms of how future transport technologies may play out for older people and ageing populations. Thinking through these issues can help us to improve research that informs our understanding of future possibilities; it can help us to prompt discussion about desirable and undesirable future directions for our societies; and it can help us to proactively develop policies and strategies that lead towards positive outcomes for all members of our communities.

References

Adey, P. 2010. *Mobility*. Abingdon, UK: Routledge.

Alessandrini, A., A. Campagna, P.D. Site, F. Filippi, and L. Persia. 2015. Automated Vehicles and the Rethinking of Mobility and Cities. *Transportation Research Procedia* 5: 145–160. https://doi.org/10.1016/j.trpro.2015.01.002.

Anderson, J.M., N. Kalra, K.D. Stanley, P. Sorensen, C. Samaras, and O. Oluwatola. 2014a. *Autonomous Vehicle Technology: A Guide for Policymakers*, 185. Santa Monica, CA: Rand Corporation.

Anderson, J.M., N. Kalra, K.D. Stanley, P. Sorensen, C. Samaras, and O. Oluwatola. 2014b. *Autonomous Vehicle Technology: How to Best Realize Its Social Benefits*, 5. Santa Monica, CA: Rand Corporation.

Bansal, P., K.M. Kockelman, and A. Singh. 2016. Assessing Public Opinions of and Interest in New Vehicle Technologies: An Austin Perspective. *Transportation Research Part C: Emerging Technologies* 67: 1–14. https://doi.org/10.1016/j.trc.2016.01.019.

Bennett, A. 1999. Subcultures or Neo-Tribes? Rethinking the Relationship Between Youth, Style and Musical Taste. *Sociology* 33 (3): 599–617.

Bergman, N., T. Schwanen, and B.K. Sovacool. 2017. Imagined People, Behaviour and Future Mobility: Insights from Visions of Electric Vehicles and Car Clubs in the United Kingdom. *Transport Policy* 59: 165–173. https://doi.org/10.1016/j.tranpol.2017.07.016.

Bloomberg Philanthropies (Cartographer). 2017. Global Atlas of Autonomous Vehicles in Cities. Retrieved from http://avsincities.bloomberg.org/global-atlas.

Blumgart, J. 2013. Whither the Driverless Car? *Next City*, January 23. Retrieved from https://nextcity.org/daily/entry/whither-the-driverless-car.

Burton, E.J., L. Mitchell, and C.B. Stride. 2011. Good Places for Ageing in Place: Development of Objective Built Environment Measures for Investigating Links with Older People's Wellbeing. *BMC Public Health* 11 (1): 839. https://doi.org/10.1186/1471-2458-11-839.

Büscher, M., J. Urry, and K. Witchger (eds.). 2011. *Mobile Methods*. Abingdon, UK: Routledge.

Butler, J. 1999. *Gender Trouble: Feminism and the Subversion of Identity*. New York, NY: Routledge.

Cavoli, C., B. Phillips, T. Cohen, and P. Jones. 2017. *Social and Behavioural Questions Associated with Automated Vehicles: A Literature Review*, 124. London, UK: Department for Transport.

Chatterjee, K., H. Sherwin, and J. Jain. 2013. Triggers for Changes in Cycling: The Role of Life Events and Modifications to the External Environment. *Journal of Transport Geography* 30: 183–193. https://doi.org/10.1016/j.jtrangeo.2013.02.007.

Clark, B., G. Parkhurst, and M. Ricci. 2016. *Understanding the Socioeconomic Adoption Scenarios for Autonomous Vehicles: A Literature Review*, 35. Bristol, UK: University of the West of England.

Clarke, D.B., M.A. Doel, and K.M.L. Housiaux (eds.). 2003. *The Consumption Reader*. London, UK: Routledge.

Cloudsley, T. 2007. After Subculture. *The European Legacy* 12 (3): 361–364. https://doi.org/10.1080/10848770701287081.

Demiris, G., M.J. Rantz, M.A. Aud, K.D. Marek, H.W. Tyrer, M. Skubic, and A.A. Hussam. 2004. Older Adults' Attitudes Towards and Perceptions of 'Smart Home' Technologies: A Pilot Study. *Medical Informatics and the Internet in Medicine* 29 (2): 87–94. https://doi.org/10.1080/14639230410001684387.

Department for Transport. 2015. *The Pathway to Driverless Cars: A Detailed Review of Regulations for Automated Vehicle Technologies*, 189. London, UK: Department for Transport.

Dowling, R., and J. Kent. 2015. Practice and Public–Private Partnerships in Sustainable Transport Governance: The Case of Car Sharing in Sydney, Australia. *Transport Policy* 40 (Suppl. C): 58–64. https://doi.org/10.1016/j.tranpol.2015.02.007.

Duncan, M., N. Charness, T. Chapin, M. Horner, L. Stevens, A. Richard, and D. Morgan. 2015. *Enhanced Mobility for Aging Populations Using Automated Vehicles*, 201. Tallahassee, FL: Florida State University.

Edwards, D. 2012. Self-driving Cars: 'Freedom' or 'More of the Same'. Retrieved from https://progressivetransit.wordpress.com/2012/02/02/self-driving-cars-freedom-or-more-of-the-same/.

Fitt, H. 2016. *The Influences of Social Meanings on Everyday Transport Practices: A Thesis Submitted in Partial Fulfilment of the Requirements for the Degree of Doctor of Philosophy in Geography at the University of Canterbury.* Doctor of Philosophy, Canterbury, New Zealand. Retrieved from http://ir.canterbury.ac.nz/bitstream/handle/10092/11846/Fitt_%20Helen%2c%20Thesis%20Final.pdf?sequence=1&isAllowed=y.

Gamson, J. 2003. Sexualities, Queer Theory, and Qualitative Research. In *The Landscape of Qualitative Research: Theories and Issues*, vol. 1, 2nd ed., ed. N.K. Denzin and Y.S. Lincoln, 540–568. Thousand Oaks, CA: Sage.

Giddens, A. 1979. *Central Problems in Social Theory: Action, Structure and Contradiction in Social Analysis.* London, UK: Macmillan.

Gleave, J. 2017. The Social Capital of Autonomous Vehicles. Retrieved from https://transportfutures.co/the-social-capital-of-autonomous-vehicles-d1949d826978.

Graves, B. 2017. Transportation and the Challenge of Future-Proofing Our Cities. Retrieved from http://www.governing.com/blogs/view/gov-transportation-challenge-future-proofing-cities.html.

Guell, C., J. Panter, N.R. Jones, and D. Ogilvie. 2012. Towards a Differentiated Understanding of Active Travel Behaviour: Using Social Theory to Explore Everyday Commuting. *Social Science and Medicine* 75 (1): 233–239. https://doi.org/10.1016/j.socscimed.2012.01.038.

Halnon, K.B., and S. Cohen. 2006. Muscles, Motorcycles & Tattoos: Gentrification in a New Frontier. *Journal of Consumer Culture* 6 (1): 33–56. https://doi.org/10.1177/1469540506062721.

Harada, T., and G. Waitt. 2013. Researching Transport Choices: The Possibilities of 'Mobile Methodologies' to Study Life-on-the-Move. *Geographical Research* 51 (2): 145–152. https://doi.org/10.1111/j.1745-5871.2012.00774.x.

Haratsis, B. 2016. *Australian Driverless Vehicle Initiative Position Paper: Economics Impacts of Automated Vehicles on Jobs and Investment.* Adelaide, Australia: Australian Driverless Vehicle Initiative.

Harper, C.D., C.T. Hendrickson, S. Mangones, and C. Samaras. 2016. Estimating Potential Increases in Travel with Autonomous Vehicles for the Non-driving, Elderly and People with Travel-Restrictive Medical Conditions. *Transportation Research Part C: Emerging Technologies* 72: 1–9. https://doi.org/10.1016/j.trc.2016.09.003.

Hawley-Hague, H., E. Boulton, A. Hall, K. Pfeiffer, and C. Todd. 2014. Older Adults' Perceptions of Technologies Aimed at Falls Prevention, Detection or Monitoring: A Systematic Review. *International Journal of Medical Informatics* 83 (6): 416–426. https://doi.org/10.1016/j.ijmedinf.2014.03.002.

KPMG. 2013. *Self-driving Cars: Are We Ready?* 36. KPMG.

Krueger, R., T.H. Rashidi, and J.M. Rose. 2016. Preferences for Shared Autonomous Vehicles. *Transportation Research Part C: Emerging Technologies* 69: 343–355. https://doi.org/10.1016/j.trc.2016.06.015.

Kyriakidis, M., R. Happee, and J.C.F. de Winter. 2015. Public Opinion on Automated Driving: Results of an International Questionnaire Among 5000 Respondents. *Transportation Research Part F: Traffic Psychology and Behaviour* 32 (Suppl. C): 127–140. https://doi.org/10.1016/j.trf.2015.04.014.

Lakey, L. 2010. *Nursing and Midwifery Council Review of Pre-registration Education: Report on the Experiences and Views of People with Dementia and Carers*, 7. London, UK: Alzheimer's Society.

Law, R. 1999. Beyond "Women and Transport": Towards New Geographies of Gender and Daily Mobility. *Progress in Human Geography* 23 (4): 567–588.

Litman, T. 2017. *Autonomous Vehicle Implementation Predictions: Implications for Transport Planning*, 24. Victoria, Canada: Victoria Transport Policy Institute.

Lyons, G., C. Davidson, T. Forster, I. Sage, J. McSaveney, E. MacDonald, A. Kole, et al. 2014. *Future Demand: How Could or Should Our Transport System Evolve in Order to Support Mobility in the Future?* 60. Wellington, NZ: Ministry of Transport.

Lyons, J.E., C.M. Swartz, H.Z. Lewis, E. Martinez, and K. Jennings. 2017. Feasibility and Acceptability of a Wearable Technology Physical Activity Intervention with Telephone Counseling for Mid-Aged and Older Adults: A Randomized Controlled Pilot Trial. *JMIR Mhealth Uhealth* 5 (3): e28. https://doi.org/10.2196/mhealth.6967.

Maffesoli, M. 1996. *The Time of the Tribes: The Decline of Individualism in Mass Society*. Trans. D. Smith. London, UK: Sage.

Mansvelt, J. 2017. *Information and Communication Technology Journeys in Later Life: From Pavements to Superhighways*. Paper Presented at the 8th New Zealand/Aotearoa Mobilities Symposium: Pavements and Paradigms: Bringing Community Back Into Mobilities, Dunedin, University of Otago.

Margolis, J. 2017. Our Utopian, Dystopian Future with Self-driving Cars. *Financial Times*, March 22. Retrieved from https://www.ft.com/content/3922742a-0d56-11e7-a88c-50ba212dce4d.

Marshall, C., and G.B. Rossman. 2011. *Designing Qualitative Research*, 5th ed., Thousand Oaks, CA: Sage.

McMahon, K.S., B. Lewis, M. Oakes, W. Guan, F.J. Wyman, and J.A. Rothman. 2016. Older Adults? Experiences Using a Commercially Available Monitor to Self-track Their Physical Activity. *JMIR mHealth uHealth* 4 (2): e35. https://doi.org/10.2196/mhealth.5120.

Middleton, J. 2009. "Stepping in Time": Walking, Time, and Space in the City. *Environment and Planning A* 41 (8): 1943–1961. https://doi.org/10.1068/a41170.

Mihailidis, A., A. Cockburn, C. Longley, and J. Boger. 2008. The Acceptability of Home Monitoring Technology Among Community-Dwelling Older Adults and Baby Boomers. *Assistive Technology* 20 (1): 1–12. https://doi.org/10.1080/10400435.2008.10131927.

Ministry of Transport. 2016. *Regulation 2025: Scenarios Summary and Key Findings*, 30. Wellington, NZ: Ministry of Transport.

Muggleton, D. 2007. Subculture. In *The Blackwell Encyclopedia of Sociology*, vol. 10, ed. G. Ritzer, 4877–4880. Malden, MA: Blackwell.

Muoio, D. 2017, September 28. Ranked: The 18 Companies Most Likely to Get Self Driving Cars on the Road First. *Business Insider Australia*.

Petroff, A. 2017. These Countries Want to Ditch Gas and Diesel Cars. *CNN.com*, July 26. Retrieved from http://money.cnn.com/2017/07/26/autos/countries-that-are-banning-gas-cars-for-electric/index.html.

Pile, S. 2010. Emotions and Affect in Recent Human Geography. *Transactions of the Institute of British Geographers* 35 (1): 5–20. https://doi.org/10.1111/j.1475-5661.2009.00368.x.

Plautz, J. 2016. Will Driverless Cars Become a Dystopian Nightmare? *The Atlantic*, January 26. Retrieved from https://www.theatlantic.com/politics/archive/2016/01/will-driverless-cars-become-a-dystopian-nightmare/459222/.

Polonetsky, J., and H. Claypool. 2016. Self-driving Cars: Transforming Mobility for the Elderly and People with Disabilities. Retrieved from https://www.huffingtonpost.com/entry/selfdriving-cars-transfor_b_12545726.html.

RAC. 2016. *Autonomous Vehicle Survey*, 4. Perth, Australia: Royal Automobile Club of WA.

Rao, M., S. Prasad, F. Adshead, and H. Tissera. 2007. The Built Environment and Health. *The Lancet* 370 (9593): 1111–1113. https://doi.org/10.1016/S0140-6736(07)61260-4.
Schwanen, T., and K. Lucas. 2011. Understanding auto motives. In *Auto Motives: Understanding Car Use Behaviours*, ed. K. Lucas, E. Blumenberg, and R. Weinberger, 3–38. Bingley, UK: Emerald.
Shaw, J., and I. Docherty. 2014. *The Transport Debate*. Bristol, UK: Policy Press.
Shergold, I., G. Lyons, and C. Hubers. 2015. Future Mobility in an Ageing Society—Where Are We Heading? *Journal of Transport & Health* 2 (1): 86–94. https://doi.org/10.1016/j.jth.2014.10.005.
Shergold, I., M. Wilson, and G. Parkhurst. 2016. *The Mobility of Older People, and the Future Role of Connected Autonomous Vehicles*, 52. Bristol, UK: Centre for Transport and Society, University of the West of England.
Shove, E., M. Pantzar, and M. Watson. 2012. *The Dynamics of Social Practice: Everyday Life and How It Changes*. London, UK: Sage.
Statistics New Zealand. 2016. *National Population Projections: 2016(Base)–2068*. Wellington, NZ: Statistics New Zealand.
The Economist. 2017. Zooming Ahead: China Moves Towards Banning the Internal Combustion Engine. *The Economist*, September 14 (Online).
United Nations, Department of Economic and Social Affairs, Population Division. 2015. *World Population Ageing 2015*, 149. New York, NY: United Nations.
Vaughan, A. 2017. All Volvo Cars to Be Electric or Hybrid from 2019. *The Guardian*, July 5. Retrieved from https://www.theguardian.com/business/2017/jul/05/volvo-cars-electric-hybrid-2019.
Wakabayashi, D. 2017. Where Driverless Cars Brake for Golf Carts. *The New York Times*, October 4. Retrieved from https://www.nytimes.com/2017/10/04/technology/driverless-cars-testing.html?smid=pl-share.
Wells, H. 1932. *Communications 1922–1932: Extract from the National Programme Radio Broadcast*. London, UK: BBC.

10

Future Ageing Populations and Policy

Judith Phillips and Shauna McGee

Introduction

Designing and implementing policy in relation to transport and mobility is both a challenge, given the increasing diversity of the ageing population in the context of rapid urbanisation, and an opportunity for creative and innovative solutions to emerge. Transport is often neglected in ageing discourse yet is a vital component in the ecosystem of policy on ageing. Despite an increasing emphasis on environmental aspects of ageing, including mobility and accessibility (Musselwhite and Haddad 2010; Rantanen 2013), and the importance of place and identity (Peace et al. 2006), transport itself

J. Phillips (✉)
University of Stirling, Stirling, UK
e-mail: judith.phillips@stir.ac.uk

S. McGee
Department of Psychology, Psychopathology and Clinical Intervention,
University of Zürich, Zürich, Switzerland
e-mail: s.mcgee@psychologie.uzh.ch

© The Author(s) 2018

A. Curl and C. Musselwhite (eds.), *Geographies of Transport and Ageing*,
https://doi.org/10.1007/978-3-319-76360-6_10

is marginalised. Yet future generations will expect that as they age innovative solutions will be in place to support their needs for continued mobility and travel. Similarly, policies related to ageing all have a transport element to them to be considered whether it is housing, family or climate change policy. Consequently, it is vital that all policy related to ageing considers the transport needs and requirements of an ageing population. This chapter looks at the broader policy challenges and opportunities going forward in designing transport for future ageing populations. We look at how future changes in policy and strategic direction could affect older people's transport and mobility needs. We argue that policy in relation to housing, social care, health, and transport need to be joined up, while policies on climate change, etc., which impact on transport and mobility, need to be age proofed.

Older People, Mobility, and Transport

Policy related to older people operates within a context to keep people active, cognitively engaged and connected to the community. The policy at EU and UK level centres around 'active ageing', 'ageing in place', and 'independent living' (Morris 2014; WHO 2002, 2007). Transport and mobility are vital to the realities of policy rhetoric.

Despite people generally being in better health, and with more opportunities to be physically fitter than ever before, those aged 65 and over are the group most likely to be physically restricted when needing to travel. For example, they are more likely than younger people to be unable to walk or cycle for extended periods of time and have more difficulty in physically accessing public transport (Schlag et al. 1996). They are also likely to be reducing the amount they drive or indeed have given up driving altogether (Box et al. 2010). Hence, older people are the most likely group to experience mobility deprivation (DfT, Department for Transport, UK 2001), and those aged 75 and over report the greatest difficulties in accessing local amenities (Age UK 2014), and engaging with and feeling part of their local community (Shergold et al. 2012).

The need to be mobile and to travel is also related to psychological well-being in older age, as a reduction in mobility can lead to an increase in isolation, loneliness, depression, or dependency on others (Fonda et al. 2001; Ling and Mannion 1995), and an overall poorer quality of life (Gabriel and Bowling 2004; Schlag et al. 1996). Giving up driving in particular has repeatedly been shown to relate to a decrease in well-being and an increase in depression and related health problems, including feelings of stress and isolation and also increased mortality (Edwards et al. 2009; Fonda et al. 2001; Ling and Mannion 1995; Marottoli and Richardson 1998; Mezuk and Rebok 2008; Musselwhite and Haddad 2010; Musselwhite and Shergold 2013; Peel et al. 2002; Ragland et al. 2005; Windsor et al. 2007; Ziegler and Schwanen 2011).

There are also societal benefits of increasing travel for older people, including the economic benefits of older people spending in shops, looking after grandchildren, undertaking voluntary work, and carrying out other caring responsibilities (WRVS 2013). Good mobility is important for enabling people to engage with their community, meet other people, and take part in social activities. Individualistic, personalised modes of transport and mobility are becoming increasingly available, and many older people are more connected than before through the digitalised world of multichannel communications.

Given this context, a challenge for any policy is how to accommodate an increasing ageing population with divergent needs and requirements for transport. Looking forward, the number of people aged 65 and over in the UK is expected to increase from 12.4 million in 2014 to 16.5 million by mid-2039, an increase of 32.7% in this short period (Office for National Statistics 2015). In addition, the population that is aged 75 and over is predicted to increase by 89.3% to 9.9 million by mid-2039, and the population aged 85 and over is predicted to more than double in the next 20 years, reaching 3.6 million by mid-2039. This means that by mid-2039, 1 in 12 people will be aged 80 or older (Office for National Statistics 2015). Against this backcloth is the challenge of mapping policy to meet the diverse needs of a heterogeneous older population in terms of wealth, disability, race, and ethnicity.

Furthermore, in a society of increasing inequality, the need to be inclusive is a growing challenge for policy.

Given the rapid changes in size and age structure of the UK population, it is essential to understand the impact and implications of these changing demographics for policy and service delivery within the context of transport.

The Diversity of Ageing

The diversity of ageing is recognised through the geography of ageing. Keating et al. (2013), drawing on the person-environment fit concept (Lawton and Nahemow 1973), argue that 'ageing in place' in policy terms should mean ageing in the right place where the environment meets the needs of the older person and can be flexible to accommodate change. They argue that those who are marginalised or frail have different needs than those who are active or volunteers. But we need to consider people in context, so, people with the same characteristics and needs may find a good fit in one community but a poor fit in another. They suggest that there is a need to move from a static concept of what constitutes 'age-friendly' to a more nuanced approach that incorporates people, place, and time. Fit also refers to the 'fit' with the history of place and 'place making', as well as with the life course of the community. Mobility and connectivity is very much part of this 'fit' which can play out in a spatial context of both rural and urban environments.

Rurality

A European study on mobility in later life found that older adults in rural areas were particularly at risk of loss of mobility in comparison with older adults in urban areas (Mollenkopf et al. 2004). The consequences of a lack of or loss in mobility disproportionately affect people living in rural and remote areas. For instance, a loss of mobility can result in reduced use of preventative services, primary care, and hospital care, due to the geographical inaccessibility of the services and the costs and inconveniences of longer journeys. This in turn can lead to worse health outcomes for older adults, in comparison with their urban counterparts (Haynes and Gale 2000). Connectivity and good transport must therefore be part of the policy on service delivery of social care, health, and housing.

Deprived Communities

Older people living in socially deprived urban neighbourhoods are also susceptible to problems such as social exclusion and decreased quality-of-life arising from the closure of local services and amenities, social polarisation, crime-related problems, poor housing, etc. (Buffel et al. 2013; Smith 2009). In relation to transport, deprived communities are often situated around main roads—people in these communities are less likely to drive, but more likely to be victims of road crashes as a pedestrian, more likely to suffer from high pollution, and have their community severed by a main road, etc. (Christie et al. 2009; Lowe et al. 2011). For older adults with restricted mobility, the loss or lack of local health and social care services, public transport, or affordable local shops can be problematic. It can necessitate the use of costly means of transport, such as taxis, and/or dependence on others to reach essential services (Scharf et al. 2001).

Ethnicity and Ethnic Minority Groups

The literature on mobility among adults from black, Asian, and minority ethnic groups in predominantly urban areas of the UK (Higgs and Gilleard 2015; Owen 2013; Rees et al. 2012) suggests that they are more likely to depend on public transport than white ethnic groups. However, fear of racial discrimination and difficulties with language represent barriers to public transport use (Smith et al. 2006). Evidence of poorer health in midlife among many members of the black and minority ethnic population suggests that this may lead to greater disability and frailty, which in turn affect social, civic, and cultural engagement (Matthews et al. 2014). Research by the Department for

Transport (2012) found that public transport providers have an inadequate understanding of the transport needs of minority ethnic groups. They are often not included in consultation and customer care surveys, and complaints procedures are often ineffective due to language difficulties. Transport must accommodate these differences and provide a non-standardised approach, ensuring an infrastructure that can be flexible and offer safety to all groups.

Minority groups in rural areas often face significant barriers to transport and mobility, particularly with growing numbers of older individuals with an increased reliance on public transport services. Looking beyond the UK, for example, the number of older Māori people, the Indigenous people of New Zealand, living in isolated rural communities is rising, increasing the demand for appropriate mobility services (Dwyer et al. 1999). Māori elders play an important role in their communities, maintaining the spiritual and cultural customs and traditions. Access to services which facilitate their continued participation in their community is therefore essential. However, a recent study on the transport system for the ageing population in the Bay of Plenty identified a number of transport issues which could affect the quality of life of this older rural minority group. This included infrastructural and design issues, integration problems between land use and transport planning, transport system issues, and the cost of transport (Bay of Plenty Regional Council 2010). Research on the Australian Indigenous population has shown similar accessibility issues and challenges. The Indigenous people in Australia are a heterogeneous group with distinct language groups, subcultures, and great social, cultural, and geographical diversity. They live in various environments including urban, rural, and remote areas, with corresponding differences in cultural identification, beliefs, and needs (Clapham and Duncan 2017). While older Indigenous people in rural and remote areas face transport barriers in accessing services, accessibility issues also effect urban areas, such as lack of adequate public transport resources, social isolation, and difficulty accessing culturally appropriate services (Arkles et al. 2010; NSW Aboriginal Transport Network 2006). A crucial issue is that care and transport policies have traditionally been developed based on government priorities rather than community needs, resulting in a lack of services for older Indigenous adults (Aboriginal and Torres Strait Islander Ageing Committee 2010; Productivity Commission 2011).

Urbanisation

A further challenge is the urbanisation of ageing. Cities and urban areas are also ageing and becoming equally diverse, consequently posing challenges for policy that is inclusive of older people (as above). Future policy requires reimaging the city for older people and the need to plan cities for living in. This is in the context of uncertainty and competing agendas for resources—to become healthier and reduce obesity, to work collaboratively within city networks, and to develop digital technology. Given that this is increasingly set within a diverse and devolved context in the UK, then increasing ways of including local older peoples' needs in policy and planning is crucial.

Given that most older people live in urban areas and will continue to do so (Moir et al. 2014; OECD 2015), exploring the urban context of transport and ageing is vital if innovative solutions are to be found to enhance quality of life.

Smart Cities

In 2011, the European Commission launched the Smart Cities and Communities European Innovation Partnership to boost the development of smart technologies in cities by pooling research resources from energy, transport, and Information and Communications Technology (ICT). These resources are concentrated on a small number of demonstration projects, which are implemented in partnership with cities (EIP-SCC 2013). The development of smart city applications will make cities more efficient and improve quality of life for their inhabitants in the future, by better-coordinating transport modes and reducing journey times, or facilitating more efficient water use, shopping, and waste recycling. Smart infrastructure systems could provide far better integration between sectors (e.g. between health and transport), which could be beneficial to older people if their needs are consciously addressed in the implementation of smart city development. Smart cities use data (often generated by intelligent sensors in a city's transport infrastructure, buildings, energy, waste, and water networks), together with information provided by residents, to make systems work together more efficiently and to help manage the city more effectively (Olofsdotter et al. 2013; Williams 2014). Consequently, a more inclusive policy incorporating the needs of older people may be realistic. This could include for example a system regulating traffic lights to enable older people to cross roads safely throughout their whole journey.

Alongside smart cities is the development of a 'Healthy Cities' network, which has engaged in implementing cycling networks, removing physical barriers to walkability, and inserting new parks into dense cities (WHO 2008). The implementation of these measures encourages health-enhancing physical activity and social cohesion, making active living, physical activity, and pedestrian mobility a core part of city development policies and plans (WHO 2009). Many such cities have become 'Supportive Environments', encouraging active socialisation and increased mobility for older adults through improved walkability, especially in areas where older people live, by providing well-maintained pavements, benches for resting, adequate lighting, and shaded and attractive streetscapes. In addition, these cities improve accessibility to public transport, provide transport to recreation facilities, and provide spaces for older people to be active in local parks and green spaces with activities such as organised group walks. In cold climates and in winter, these cities ensure that sidewalks are clear of ice (Edwards and Tsouros 2008).

Nevertheless, very few policy areas, particularly around smart cities, have placed older people's needs in the forefront, or considered social and health care delivery as part of their transport ecosystem. However, the opportunity for joining up these initiatives has come via the age-friendly cities and communities' movement, which attempts to nurture caring and supportive environments.

Since 2007, when the World Health Organization (WHO) Global Network of Age-Friendly Cities and Communities was established, policies have been developed around age-friendly cities and more recently, dementia supportive communities (Alzheimer's Society 2013).

Age-friendly cities have introduced policies and holistic action plans addressing the health needs of older people that emphasise participation, empowerment, independent living, supportive and secure physical and social environments, and accessible services and support (WHO 2009). Creating resilient and supportive environments is a core theme in policy priorities for health and well-being particularly in the EU. Cities are encouraged to improve urban planning to enhance the mobility of ageing populations or people with disabilities to improve their health and well-being.

Examples of this approach in the UK include Belfast (links & walkability assessment; Belfast Healthy Cities 2014) and Manchester (transport and age-friendly development plan). Manchester's healthy ageing programme Valuing Older People (VOP), now known as the Age-Friendly Manchester (AFM) Programme, was established in 2003 to facilitate engagement with older people and build a consideration of both their needs and potential contribution into a wide range of the city's strategies (Manchester City Council 2009; McGarry and Morris 2011). The VOP board ensured that the voice of older people was included in transport plans across the city, and by 2008, the VOP update report had identified many achievements addressing older people's mobility needs, including free off-peak travel on local buses, trains, and trams for over-60s and the participation of taxis in a travel voucher scheme providing substantial discount to older adults. Since the first VOP strategy was launched in 2004, many of the city's agencies have contributed to the development of structures which improve the lives of older adults. For example, £60 million was invested in the built environment to improve bus corridors and make walking facilities and areas around bus stops safer, for instance, by raising the pavements to make it easier for older adults with mobility issues to board and disembark the bus (Valuing Older People 2008). In 2009, the AFM partnership produced a strategy for ageing, an initiative to improve life for older people in Manchester which involved many different services, organisations, agencies, and older Manchester residents. This strategy expands on the work done by VOP to ensure older adults' mobility needs are addressed, enabling them to retain their independence and access support and services when needed (Edmondson and Von Kondratowitz 2009).

Recently the AFM programme developed a work plan for 2016 and 2017, which identified objectives to improve the lives of older people in Manchester, working in cooperation with older people and organisations in the public, private, voluntary, and community sectors. The intended outcome of the work plan is to create and provide age-friendly neighbourhoods and services which improve accessibility options for older adults, and to improve the involvement and communication of older adults in their community. It further aims to combine research, policy, and practice to increase the recognition and implementation of age-friendly practices. More specifically, one objective set out to promote the needs of older people in the design and development of new transport infrastructure and services (Manchester City Council 2016). In line with this, in 2017 the AFM Older People's Board provided a formal contribution to the Transport for Greater Manchester Vision 2040 plan and strategy, which aims to provide better transport networks and promote independence and active lifestyles (Manchester City Council 2017; Transport for Greater Manchester 2017). The achievements of the VOP and AFM programmes highlight the power of involving older people and their perspectives in the design and development of city- and transport-related strategies. Going forwards, older people should be involved in the decision-making processes from an early stage to help make policy development and implementation as encompassing and as effective as possible.

In a 'Research and Evaluation Framework' for age-friendly cities, Handler (2014) emphasises the importance of having a good social and emotional fabric of a city. This highlights the importance of people's subjective connections with place, the meanings and values attached to places (Phillips et al. 2011), and the perceptions of an environment which can give people confidence in a place. Good transport is part of creating and maintaining that confidence.

One critique of the age-friendly cities is that it often misses the emotional aspects, focusing on improving the fabric of the built environment. UK policy has embraced many of the above concepts through '*Lifetime Homes: Lifetime Neighbourhoods*'. However, the lack of policy uptake on this at a local level has resulted in a series of initiatives which are fragmentary and disconnected. For instance, local planning needs assessment often do not consider older people's transport needs related to where they live or work; social care policy is based on budgets and integration with health services. Peoples' relationship with transport networks, systems, and connectivity is fundamental to how they experience and perceive quality of life and assess place. However, place as a policy concept is gaining traction through with a focus on inclusivity.

Global policies such as age-friendly cities are gathering momentum, and despite recessionary cutbacks are being taken forward by local authorities, along with local projects such as dementia-friendly communities.

Age-friendly communities and lifetime neighbourhoods may provide the environmental context/concepts for ageing policy to draw on, yet there will be considerable future challenges to policy enactment.

Future of Transport for Older Adults: What Will Future UK Policy Need to Address Given the Diversity of Older People and Cities?

In 2013, 50–59-year-olds drove the highest number of miles per person per year, closely followed by 40–49-year-olds. They spent a less than average number of miles walking and using buses (DfT, Department for Transport 2014). In 20 years, many in these age groups will be reaching retirement age and may have different expectations based around a personal motorised future, whether through active retirement or through continuing work. This will be coupled with increasing technologies placed within the vehicle to aid driving. Driverless cars may have increased personal safety for the older driver, who is more vulnerable in a collision, but also for other road and pavement users generally. Autonomous vehicles may also have an impact on prolonging and enhancing the ability to travel and enabling people to feel safe (Musselwhite 2011; Shergold et al. 2015).

More personal mobility among 65-year-olds will apply for both males and females compared to previous generations, but again females are more likely to be driven by others and this is set to continue. Technology that is currently not seen as economically viable, such as hydrogen power, could have reduced in cost and so provide other means to continue with the predominance of cars, rather than moving people towards other modes of transport (Ball and Wietschel 2009).

Training courses will have to be introduced more widely to provide a variety of mobility options, especially for those who may be using public transport for the first time since their teenage years (Ormerod et al. 2015). One such solution is 'travel training', which provides practical help to individuals travelling by public transport, by foot, or by bicycle. Travel training can be supported and delivered by a variety of (often collaborating) organisations, including local authorities, transport departments, voluntary organisations, and even schools and employment agencies (DfT, Department for Transport 2011). One example of such a travel training service is the 'bus buddies' scheme in which volunteers assist older adults in using buses in their local area. A number of towns in the UK have successfully set up this scheme in collaboration with volunteer centres (Osborne n.d.; Ribble Valley Borough Council 2015). There will also be some stigma attached to this, especially given that the decline in bus services is likely to continue and there will be more community transport provision. Buses, still free for those aged over 68, are likely to be confined to principle routes in town and city areas, with suburbs and rural areas not being served. Community transport will take the place of some of the services in these areas, but is likely to be confined to a small minority of very frail passengers, to which it is a lifeline.

There will be a widening gap between older people who are very fit and well, being potentially more affluent, and those with chronic impairments that allow for use of concessions, and who are eligible for community transport. This introduces an increase in a 'squeezed middle', whose health and ability to navigate driving, walking, or using public transport, may vary from day to day. Mobility options for such a group are hard to meet.

The advent of technology, coupled with changes in distribution of wealth, may mean car ownership is more likely to be shared, along with the use of paratransit, such as automated taxis with a supervisor controlling many vehicles at once. Hence, personal mobility for older people especially out of town centres will be maintained, especially if they are able to share the cost of ownership around a community. Larger automated vehicles may serve city and town centres mimicking automated buses. Technology to command a personalised service could exclude unless rolled out across all areas. Technology however will dictate to some degree how and where we live, how we support ourselves, and how we engage with others in later life.

Given the above scenario of changing mobility needs and transport how will policy respond?

1. We need to challenge concepts enshrined in policy in general such as 'ageing in place' and 'remaining independent'. Just being in a place where the environment allows you to age in place does not necessarily mean it is ideal or the right 'fit'—it can mean loss of social networks and poor accessibility options. The focus of policy has been on the location, design, and built environmental context of ageing in place. This has been reinforced through the policy initiatives to future-proof design and transport systems so they don't exclude. Travel and transport is mostly interdependent, and as we age within a place such interdependency (on other road users, for care, etc.) needs to be factored into policy options. The social and relational aspects of space and place are critical in the design of our walking and driving environment. Bringing planners, urban designers, architects, and engineers together with older people to design safe public space with good transport links that encourage interaction is critical to quality of life.
2. A future of inequality—transport policy itself can lead to a spatial concentration of disadvantage as roads and routes cut through communities separating the rich from areas of deprivation and poor infrastructure (Harvey 2009). This could be exaggerated through smart technologies and solutions only being available to certain areas in the city.
3. The above assumes that place is intergenerational, bringing together all age groups and with housing policy geared to integrated streets for life and offering choice. Building for co-housing (e.g. UK Co-housing Network 2017) and similar alternatives (e.g. housing above retail stores in the high street, enabling older people direct

access to services) that reduce the need for reliance on transport can also address other policy and planning issues such as revitalising town centres. The wider land use policy context is critical to future transport and mobility considerations with aspects of lighting, green spaces, transport routes, nodes, networks, location of toilets, and development of safer roads and infrastructure part of the housing mix. While housing of 'extra care' and 'lifetime homes' has much to offer it can exclude the majority of older people; as the majority of older people live in non-specialist housing, there is a need to offer choice of ordinary housing options that connect with transport routes to central facilities and services.

4. Support and funding to enable lifelong mobility may also be part of the intergenerational contract. Policies earlier in life will have an impact on whether use of car or public transport is preferred. Changing behaviour to ensure a healthier commute to shops and work for example strays into health policy yet requires provision of suitable options to the private car as well as educational campaigns to promote maximum mobility and support for people to continue to drive and walk safely. Improving accessibility so people can participate in their community is a vital health and quality-of-life consideration.
5. Work and retirement presents a changing landscape for ageing with many older people working longer (Mermin et al. 2007; Scherger 2015). The needs of older commuters for safer, better-designed vehicles, better greener transport routes, and flexible transport options and costs will need to be factored into policy at many levels.
6. Social care policy also needs to address how care networks rapidly connect if provision of care at home provided in the 'informal' sector is to be effective. Personalised transport and mobility options (e.g. smart options at bus stops, taxis from home) need to be more available.
7. Climate change policy and energy efficiency will increasingly look to transport policy for solutions to cut emissions and deliver a greener future. Older people as drivers may need to recognise their

responsibilities in this policy arena in their leisure and working lives. Climate change policy needs to be age proofed if it is to deliver. Older people as a social group are often overlooked as potential change agents for mitigating climate change. However, research suggests that baby boomers (people aged 50–65) in high-income countries, such as the UK, have the largest carbon footprint (Haq et al. 2010) whilst, significantly, older age groups are among the most sceptical about the reality of climate change, least concerned about its effects, and least engaged behaviourally (Capstick et al. 2013; Whitmarsh et al. 2011; Poortinga et al. 2011).

8. Active ageing and active citizenship at the heart of policy requires the involvement of older people in policy development. A city's own sense of a liveable community or a good place to grow old is important—it needs to be owned as a vision by its citizens, irrespective of age (Buffel et al. 2012). Buffel and colleagues argue for older people's 'right to the city', i.e. their right to lead a life of dignity and independence, and to participate in social and cultural life (Harvey 2009; Rémillard-Boilard et al. 2017). The co-production of transport plans and solutions can ensure age-friendly routes and inclusion.

Conclusion

Given the wider policy context in which transport operates, there will be several challenges for future transport and mobility planning and policy with regard to older people. Such challenges will be:

- to join up policies—both horizontally (across health, housing, transport, and planning) and vertically between UK, EU, and globally through such initiatives, e.g. age-friendly communities;
- to become more holistic with transport, taking a complete system from door-to-door approach;
- to involve older adults in the decision-making processes from an early stage;
- to recognise individual differences and to tailor mobility interventions;

- to recognise the impact of social policy on mobility; and
- to keep people active, cognitively engaged, and healthy, as part of a wider picture.

Future cities and communities need to be resilient, inclusive, authentic, and diverse; they need to engage in new models of development which encourage mobility but do not segregate people by transport routes and corridors, and allow for participation. Some would argue (Jakob 2010) that we need new models of planning and new paradigms of place with a need for more social equity, tolerance, and trust. For instance, a recently developed concept of 'Transport Justice' proposes a distribution approach for the provision of adequate transportation based on the principles of social justice. This model focuses on the individuals who are using (or not using) the transport system, rather than focusing on the transport system itself (Martens 2012, 2017). Nevertheless, these models must be considered within the context of planning for uncertainty, where we must challenge the direction and staying power of policy to concepts such as 'ageing in place'. To future-proof policies that consider the needs of older people with uncertainty in mind will require a revolutionary change in our way of thinking about older people within the wider environmental context.

References

Aboriginal and Torres Strait Islander Ageing Committee. 2010. *Aboriginal Ageing: Growing Old in Aboriginal Communities. Linking Services and Research.* A Report of the 2nd National Workshop of the Australian Association of Gerontology (AAG) Aboriginal and Torres Strait Islander Ageing Committee (ATSIAC), held in Darwin, Northern Territory, 11 August 2010. Retrieved from https://www.aag.asn.au/documents/item/1438.

Age UK. 2014. *Agenda for Later Life 2014: Public Policy for Later Life*. London: Age UK.

Alzheimer's Society. 2013. *Building Dementia-Friendly Communities: A Priority for Everyone*. London: Alzheimer's Society.

Arkles, R., L.J. Pulver, H. Robertson, B. Draper, S. Chalkley, and G. Broe. 2010. *Ageing, Cognition and Dementia in Australian Aboriginal and Torres Strait Islander Peoples: A Life Cycle Approach. A Review of the Literature.* Sydney: Neuroscience Research Australia and Muru Marri Indigenous Health Unit, University of New South Wales.

Ball, M., and M. Wietschel. 2009. The Future of Hydrogen—Opportunities and Challenges. *International Journal of Hydrogen Energy* 34 (2): 615–627. https://doi.org/10.1016/j.ijhydene.2008.11.014.

Bay of Plenty Regional Council. 2010. *Transportation Publication 2010/03: Study of the Relationship Between an Ageing Population and the Transport System in the Bay of Plenty.* Bay of Plenty Regional Land Transport Strategy Supporting Paper No. 07. Whakatāne, New Zealand: Bay of Plenty Regional Council.

Belfast Healthy Cities. 2014. *Walkability Assessment for Healthy Ageing.* Belfast: Belfast Healthy Cities.

Box, E., J. Gandolfi, and C.G.B. Mitchell (eds.). 2010. *Maintaining Safe Mobility for the Ageing Population: The Role of the Private Car.* London: RAC Foundation.

Buffel, T., C. Phillipson, and T. Scharf. 2013. Ageing in Urban Environments: Developing 'Age-Friendly' Cities. *Critical Social Policy* 32 (4): 597–617. https://doi.org/10.1177/0261018311430457.

Buffel, T., D. Verté, L. De Donder, N. De Witte, S. Dury, T. Vanwing, and A. Bolsenbroek. 2012. Theorising the Relationship Between Older People and Their Immediate Social Living Environment. *International Journal of Lifelong Education* 31 (1): 13–32. https://doi.org/10.1080/02601370.2012.636577.

Buffel, T., P. McGarry, C. Phillipson, L. De Donder, S. Dury, N. De Witte, A.S. Smetcoren, and D. Verté. 2014. Developing Age-Friendly Cities: Case Studies from Brussels and Manchester and Implications for Policy and Practice. *Journal of Aging & Social Policy* 26 (1–2): 52–72. https://doi.org/10.1080/08959420.2014.855043.

Capstick, S., N. Pidgeon, and M. Whitehead. 2013. *Public Perceptions of Climate Change in Wales. Summary Findings of a Survey of the Welsh Public Conducted During November and December 2012.* http://c3wales.org/thematic_clusters/survey-findings-reveal-public-perceptions-of-climate-change/. Accessed 6 October 2017.

Christie, N., H. Ward, R. Kimberlee, R. Lyons, E. Towner, M. Hayes, S. Robertson, S. Rana, and M. Brussoni. 2009. *Road Traffic Injury Risk in*

Disadvantaged Communities: Evaluation of the Neighbourhood Road Safety Initiative. Road Safety Web Publication No. 19. London: Department for Transport.

Clapham, K., and C. Duncan. 2017. Indigenous Australians and Ageing: Responding to Diversity in Policy and Practice. In *Ageing in Australia: Challenges and Opportunities*, ed. K. O'Loughlin, C. Browning, and H. Kendig, 103–126. New York: Springer Science+Business Media.

DfT (Department for Transport, UK). 2001. *Older Drivers: A Literature Review.* London: Department for Transport.

DfT (Department for Transport, UK). 2011. *Travel Training: Good Practice Guidance.* London: Department for Transport. Retrieved from https://www.gov.uk/government/uploads/system/uploads/attachment_data/file/4482/guidance.pdf.

DfT (Department for Transport, UK). 2012. *Transport for Everyone: An Action Plan to Promote Equality.* London: Department for Transport. Retrieved from https://www.gov.uk/government/uploads/system/uploads/attachment_data/file/36211/equality-action-plan.pdf.

DfT (Department for Transport, UK). 2014. *Transport Statistics Great Britain: 2013.* London: Department for Transport. Retrieved from www.gov.uk/government/uploads/system/uploads/attachment_data/file/264679/tsgb-2013.pdf.

Dwyer, M., A. Gray, and M. Renwick. 1999. *Factors Affecting the Ability of Older People to Live Independently.* Ministry of Social Policy.

Edmondson, R., and H. Von Kondratowitz (eds.). 2009. *Valuing Older People: A Humanist Approach to Ageing.* Bristol, UK: Policy Press at the University of Bristol.

Edwards, J.D., M. Perkins, L.A. Ross, and S.L. Reynolds. 2009. Driving Status and Three-Year Mortality Among Community-Dwelling Older Adults. *The Journals of Gerontology, Series A: Biological Sciences and Medical Sciences* 64 (2): 300–305.

Edwards, P., and A.D. Tsouros. 2008. *A Healthy City is an Active City: A Physical Activity Planning Guide.* Copenhagen, Denmark: World Health Organization.

EIP-SCC (European Innovation Partnership on Smart Cities and Communities). 2013. *Strategic Implementation Plan.* Retrieved from http://ec.europa.eu/eip/smartcities/files/sip_final_en.pdf.

Fonda, S.J., R.B. Wallace, and A.R. Herzog. 2001. Changes in Driving Patterns and Worsening Depressive Symptoms Among Older Adults.

The Journals of Gerontology, Series B: Psychological Sciences and Social Sciences 56 (6): S343–S351.

Gabriel, Z., and A. Bowling. 2004. Quality of Life from the Perspectives of Older People. *Ageing & Society* 24 (5): 675–691. https://doi.org/10.1017/S0144686X03001582.

Handler, S. 2014. *A Research & Evaluation Framework for Age-Friendly Cities.* UK Urban Ageing Consortium. Retrieved from https://www.housinglin.org.uk/_assets/Resources/Housing/OtherOrganisation/A_Research_and_Evaluation_Framework_for_Age-friendly_Cities.pdf.

Haq, G., D. Brown, and Hards, S. 2010. *Older People and Climate Change: The Case for Better Engagement.* http://www.seinternational.org/mediamanager/documents/Publications/Climate-mitigation-adaptation/pr%20-%20old%20people%20and%20climate%20change%20pr%20100826lowres.pdf. Accessed 6 October 2017.

Harvey, D. 2009. *Social Justice and the City.* Athens, GA: University of Georgia Press.

Haynes, R., and S. Gale. 2000. Deprivation and Poor Health in Rural Areas: Inequalities Hidden by Averages. *Health and Place* 6 (4): 275–285. https://doi.org/10.1016/S1353-8292(00)00009-5.

Higgs, P., and C. Gilleard. 2015. *Rethinking Old Age: Theorising the Fourth Age.* London: Palgrave Macmillan.

Jakob, D. 2010. Constructing the Creative Neighborhood: Hopes and Limitations of Creative City Policies in Berlin. *City, Culture and Society* 1 (4): 193–198. https://doi.org/10.1016/j.ccs.2011.01.005.

Keating, N., J. Eales, and J. Phillips. 2013. Age-Friendly Rural Communities: Conceptualizing 'Best-Fit'. *Canadian Journal on Aging* 32 (4): 319–332. https://doi.org/10.1017/S0714980813000408.

Lawton, M.P., and L. Nahemow. 1973. Ecology and the Aging Process. In *The Psychology of Adult Development and Aging*, ed. C. Eisdorfer and M.P. Lawton, 619–674. Washington, DC: American Psychological Association.

Ling, D. J., and R. Mannion. 1995. Enhanced Mobility and Quality of Life of Older People: Assessment of Economic and Social Benefits of Dial-a-Ride Services. In *Proceedings of the Seventh International Conference on Transport and Mobility for Older and Disabled People*, vol. 1. DETR, London.

Lowe, C., G. Whitfield, L. Sutton, and J. Hardin. 2011. Road User Safety and Disadvantage Road Safety Research Report No. 123. London: Department for Transport.

Manchester City Council. 2009. *Manchester: A Great Place to Grow Older 2010–2020*. Manchester City Council. Retrieved from http://www.manchester.gov.uk/downloads/file/11899/manchester_a_great_place_to_grow_.
Manchester City Council. 2016. *Age-Friendly Manchester (AFM) Work Plan 2016–17*. Manchester City Council. Retrieved from http://www.manchester.gov.uk/downloads/download/6371/age-friendly_manchester_afm_development_plan_2014-16.
Manchester City Council. 2017. Item 7: Age-Friendly Manchester Strategy. *Communities and Equalities Scrutiny Committee Meeting*. Manchester City Council. Retrieved from http://www.manchester.gov.uk/meetings/meeting/2883/communities_and_equalities_scrutiny_committee/attachment/22333.
Marottoli, R.A., and E.D. Richardson. 1998. Confidence in, and Self-Rating of, Driving Ability Among Older Drivers. *Accident Analysis and Prevention* 30 (3): 331–333. https://doi.org/10.1016/S0001-4575(97)00100-0.
Martens, K. 2012. Justice in Transport as Justice in Accessibility: Applying Walzer's 'Spheres of Justice' to the Transport Sector. *Transportation* 39 (6): 1035–1053. https://doi.org/10.1007/s11116-012-9388-7.
Martens, K. 2017. *Transport Justice: Designing Fair Transportation Systems*. New York: Routledge.
Matthews, K., P. Demakakos, J. Nazroo, and A. Shankar. 2014. The Evolution of Lifestyles in Older Age in England. In *The Dynamics of Ageing: Evidence from the English Longitudinal Study of Ageing 2002–2012*, ed. J. Banks, J. Nazroo, and A. Steptoe, 40–63. London: The Institute for Fiscal Studies.
McGarry, P., and J. Morris. 2011. A Great Place to Grow Older: A Case Study of How Manchester is Developing an Age-Friendly City. *Working with Older People* 15 (1): 38–46. https://doi.org/10.5042/wwop.2011.0119.
Mermin, G.B.T., R.W. Johnson, and D.P. Murphy. 2007. Why Do Boomers Plan to Work Longer? *The Journals of Gerontology, Series B: Psychological Sciences and Social Sciences* 62B (5): S286–S294. https://doi.org/10.1093/geronb/62.5.S286.
Mezuk, B., and G.W. Rebok. 2008. Social Integration and Social Support Among Older Adults Following Driving Cessation. *The Journals of Gerontology, Series B: Psychological Sciences and Social Sciences* 63B (5): S298–S303. https://doi.org/10.1093/geronb/63.5.S298.
Moir, E., T. Moonen, and G. Clark. 2014. *The Future of Cities: What is the Global Agenda?* London: Government Office for Science. Retrieved from https://www.gov.uk/government/uploads/system/uploads/attachment_data/file/429125/future-cities-global-agenda.pdf.

Mollenkopf, H., F. Marcellini, I. Ruoppila, Z. Széman, M. Tacken, and H.-W. Wahl. 2004. Social and Behavioural Science Perspectives on Out-of-Home Mobility in Later Life: Findings from the European Project MOBILATE. *European Journal of Ageing* 1 (1): 45–53. https://doi.org/10.1007/s10433-004-0004-3.

Morris, J. 2014. *Independent Living Strategy: A Review of Progress*. London: Disability Rights UK. Retrieved from https://www.disabilityrightsuk.org/sites/default/files/pdf/IndependentLivingStrategy-A%20review%20of%20progress.pdf.

Musselwhite, C.B.A. 2011. *Successfully Giving Up Driving for Older People.* London: The International Longevity Centre – UK. Retrieved from www.ilcuk.org.uk/files/Successfully_giving_up_driving_for_older_people_1.pdf.

Musselwhite, C.B.A., and H. Haddad. 2010. Mobility, Accessibility and Quality of Later Life. *Quality in Ageing and Older Adults* 11 (1): 25–37. https://doi.org/10.5042/qiaoa.2010.0153.

Musselwhite, C.B.A., and I. Shergold. 2013. Examining the Process of Driving Cessation in Later Life. *European Journal of Ageing* 10 (2): 89–100. https://doi.org/10.1007/s10433-012-0252-6.

NSW Aboriginal Transport Network. 2006. *Transport Disadvantage in Aboriginal Communities*. NSW: Northern Rivers Social Development Council.

OECD. 2015. *Ageing in Cities: Policy Highlights.* Paris: OECD Publishing. Retrieved from http://www.oecd.org/cfe/regional-policy/policy-brief-ageing-in-cities.pdf.

Office for National Statistics. 2015. *National Population Projections: 2014-Based Statistical Bulletin.* Office for National Statistics. Retrieved from https://www.ons.gov.uk/peoplepopulationandcommunity/populationandmigration/populationprojections/bulletins/nationalpopulationprojections/2015-10-29.

Olofsdotter, B., K. Björnberg, H.-W. Chang, J.-H. Kain, E. Linn, and B. Scurell. 2013. *Competing for Urban Land: Synthesis Report.* Research Project FP7, Urban-Nexus.

Ormerod, M., R. Newton, J. Phillips, C. Musselwhite, S. McGee, and R. Russell. 2015. *How Can Transport Provision and Associated Built Environment Infrastructure be Enhanced and Developed to Support the Mobility Needs of Individuals as the Age?* Foresight, Government Office for Science: Future of Ageing. Retrieved from https://www.gov.uk/government/publications/future-of-ageing-transport-and-mobility.

Osborne. n.d. *Bus Buddies to Help Elderly Get 'On the Buses'*. Retrieved from http://www.osborne.co.uk/2016/09/08/bus-buddies-help-elderly-get-buses/.

Owen, D. 2013. DR18: Is There Evidence that National, and Other, Identities Change as the Ethnic Composition of the UK Changes? If so, How? Are There Differences Between People from Different Ethnic Groups? And of Different Ages/Social Classes? *Future Identities: Changing Identities in the UK—The Next 10 Years*. London: Government Office for Science. Retrieved from https://www2.warwick.ac.uk/fac/soc/ier/publications/2013/do_13-520-national-identities-change-as-ethnic-composition-changes.pdf.

Peace, S., C. Holland, and L. Kellaher. 2006. *Environment and Identity in Later Life*. Maidenhead: Open University Press.

Peel, N., J. Westmoreland, and M. Steinberg. 2002. Transport Safety for Older People: A Study of Their Experiences, Perceptions and Management Needs. *Injury Control & Safety Promotion* 9: 19–24.

Phillips, J., N. Walford, and A. Hockey. 2011. How Do Unfamiliar Environments Convey Meaning to Older People? Urban Dimensions of Placelessness and Attachment. *International Journal of Ageing and Later Life* 6 (2): 73–102. https://doi.org/10.3384/ijal.1652-8670.116273.

Poortinga, W., A. Spence, L. Whitmarsh, S. Capstick, and N.F. Pidgeon. 2011. Uncertain Climate: An Investigation into Public Scepticism About Anthropogenic Climate Change. *Global Environmental Change* 21 (3): 1015–1024. https://doi.org/10.1016/j.gloenvcha.2011.03.001.

Productivity Commission. 2011. *Caring for Older Australians* (Report No. 53). Final Inquiry Report. Canberra, Australia. Retrieved from http://www.pc.gov.au/inquiries/completed/aged-care/report/aged-care-overview-booklet.pdf.

Ragland, D.R., W.A. Satariano, and K.E. MacLeod. 2005. Driving Cessation and Increased Depressive Symptoms. *The Journals of Gerontology, Series A: Biological Sciences and Medical Sciences* 60 (3): 399–403.

Rantanen, T. 2013. Promoting Mobility in Older People. *Journal of Preventive Medicine and Public Health* 46 (Suppl 1): S50–S54. https://doi.org/10.3961/jpmph.2013.46.S.S50.

Rees, P., P. Wohland, P. Norman, and P. Boden. 2012. Ethnic Population Projections for the UK, 2001–2051. *Journal of Population Research* 29 (1): 45–89. https://doi.org/10.1007/s12546-011-9076-z.

Rémillard-Boilard, S., T. Buffel, and C. Phillipson. 2017. Involving Older Residents in Age-Friendly Developments: From Information to Coproduction Mechanisms. *Journal of Housing For the Elderly* 31 (2): 146–159. https://doi.org/10.1080/02763893.2017.1309932.

Ribble Valley Borough Council. 2015. *Best Feet Forward for Bus Buddies.* Retrieved from https://www.ribblevalley.gov.uk/news/article/622/best_feet_forward_for_bus_buddies.

Scharf, T., C. Phillipson, P. Kingston, and A.E. Smith. 2001. Social Exclusion and Older People: Exploring the Connections. *Education and Ageing* 16 (3): 303–320.

Scherger, S. (ed.). 2015. *Paid Work Beyond Pension Age: Comparative Perspectives.* London, UK: Palgrave Macmillan.

Schlag, B., U. Schwenkhagen, and U. Trankle. 1996. Transportation for the Elderly: Towards a User-Friendly Combination of Private and Public Transport. *IATSS Research* 20 (1): 75–82.

Shergold, I., G. Parkhurst, and C.B.A. Musselwhite. 2012. Rural Car Dependence: An Emerging Barrier to Community Activity for Older People? *Transportation Planning and Technology* 35 (1): 69–85.

Shergold, I., G. Lyons, and C. Hubers. 2015. Future Mobility in an Ageing Society—Where Are We Heading? *Journal of Transport & Health* 2 (1): 86–94. https://doi.org/10.1016/j.jth.2014.10.005.

Smith, A.E. 2009. *Ageing in Urban Neighbourhoods: Place Attachment and Social Exclusion.* Bristol: Policy Press at the University of Bristol.

Smith, N., J. Beckhelling, A. Ivaldi, K. Kellard, A. Sandu, and C. Tarrant. 2006. *Evidence Based Review on Mobility: Choices and Barriers for Different Social Groups.* London: Department for Transport.

Transport for Greater Manchester. 2017. *Greater Manchester Transport Strategy 2040: A Sustainable Urban Mobility Plan for the Future.* Transport for Greater Manchester: Greater Manchester Combined Authority. Retrieved from http://www.tfgm.com/2040/Pages/strategy/assets/2017/2-17-0078-GM-2040-Full-Strategy-Document.pdf.

UK Cohousing Network. 2017. *UK Cohousing. Cohousing in the UK.* Retrieved from https://cohousing.org.uk/about/cohousing-in-the-uk/.

Valuing Older People (VOP). 2008. *Update Report 2001–08.* Manchester City Council.

Whitmarsh, L. 2011. Scepticism and Uncertainty About Climate Change: Dimensions, Determinants and Change Over Time. *Global Environmental Change* 21 (2): 690–700. https://doi.org/10.1016/j.gloenvcha.2011.01.016.

Williams, K. 2014. Urban Form and Infrastructure: A Morphological Review, Future of Cities. Working Paper, London: Foresight, Government Office for Science. Retrieved from https://www.gov.uk/government/uploads/system/uploads/attachment_data/file/324161/14-808-urban-form-and-infrastructure-1.pdf.

Windsor, T.D., K.J. Anstey, P. Butterworth, M.A. Luszcz, and G.R. Andrews. 2007. The Role of Perceived Control in Explaining Depressive Symptoms Associated with Driving Cessation in a Longitudinal Study. *The Gerontologist* 47 (2): 215–223.

World Health Organization (WHO). 2002. *Active Ageing: A Policy Framework*. Geneva: World Health Organization.

World Health Organization (WHO). 2007. *Global Age-Friendly Cities: A Guide*. Geneva: World Health Organization.

World Health Organization (WHO). 2008. *City Leadership for Health: Summary Evaluation of Phase IV of the WHO European Healthy Cities Network*. Copenhagen, Denmark: WHO Regional Office for Europe.

World Health Organization (WHO). 2009. *Phase V (2009–2013) of the WHO European Healthy Cities Network: Goals and Requirements*. Copenhagen, Denmark: WHO Regional Office for Europe.

WRVS. 2013. *Going Nowhere Fast. Impact of Inaccessible Public Transport on Wellbeing and Social Connectedness of Older People in Great Britain*. Cardiff: WRVS. Retrieved from https://www.royalvoluntaryservice.org.uk/Uploads/Documents/Reports%20and%20Reviews/Trans%20report_GB_web_v1.pdf.

Ziegler, F., and T. Schwanen. 2011. 'I Like to Go Out to Be Energised by Different People': An Exploratory Analysis of Mobility and Wellbeing in Later Life. *Ageing & Society* 31 (5): 758–781. https://doi.org/10.1017/S0144686X10000498.

Index

© The Editor(s) (if applicable) and The Author(s) 2018
A. Curl and C. Musselwhite (eds.), *Geographies of Transport and Ageing*,
https://doi.org/10.1007/978-3-319-76360-6

CPSIA information can be obtained
at www.ICGtesting.com
Printed in the USA
LVOW13*1946030518
575867LV00017B/345/P